To
SANDY, DANIELLE, BECCA, AND CHELSEA

CONTENTS

PREFACE

Radio frequency identification (RFID) technology is a wireless communication technology that enables users to uniquely identify tagged objects or people.

RFID is rapidly becoming a cost-effective technology. This is in large part due to the efforts of Wal-Mart and the Department of Defense (DoD) to incorporate RFID technology into their supply chains. In 2003, with the aim of enabling pallet-level tracking of inventory, Wal-Mart issued an RFID mandate requiring its top suppliers to begin tagging pallets and cases, with Electronic Product Code (EPC) labels. The DoD quickly followed suit and issued the same mandate to its top 100 suppliers. This drive to incorporate RFID technology into their supply chains is motivated by the increased shipping, receiving and stocking efficiency and the decreased costs of labor, storage, and product loss that pallet-level visibility of inventory can offer.

Wal-Mart and the DoD are, respectively, the world's largest retailer and the world's largest supply chain operator. Due to the combined size of their operations, the RFID mandates are spurring growth in the RFID industry and bringing this emerging technology into the mainstream. The costs of employing RFID are falling as a result of the mandates also, as an economy of scale is realized. Lastly, the mandates appear to have united the industry behind a single technology standard (EPCglobal's Electronic Product Code standard). The lack of industry consensus over the standards issue had been impeding industry growth prior to the issuance of the mandates.

Wal-Mart and DoD alone cannot account for all the current interest in RFID technology, however. Given the following forecasts of industry growth, it becomes clear why RFID has begun to attract the notice of a wide range of industries and government agencies:

1. In the past 50 years, only 1.5 billion RFID tags were sold worldwide. Sales for 2004 alone are expected to top 1 billion, and as many as 1 trillion tags could be delivered by 2015.
2. Wal-Mart's top 100 suppliers alone could account for 1 billion tags sold annually.
3. Revenues for the RFID industry are expected to hit $7.5 billion by 2006.
4. Early adopters of RFID technology were able to lower supply chain costs by 3–5% and simultaneously increase revenue by 2–7% according to a study by AMR Research.
5. For the pharmaceutical industry alone, RFID-based solutions are predicted to save more than $9 billion by 2007.
6. In the retailing sector, item-level tagging could begin in five years.

In short, the use of RFID technology is expected to grow significantly in the next five years, and it is predicted that someday RFID tags will be as pervasive as bar codes.

This book provides a broad overview and guide to RFID technology and its application. It is an effort to do the initial "homework" for the reader interested in better understanding RFID tools. It is written to provide an introduction for business leaders, supply chain improvement advocates, and technologists to help them adopt RFID tools for their unique applications, and provide the basic information for better understanding RFID.

The book describes and addresses the following:

- How RFID works, how it's used, and who is using it
- The history of RFID technology, the current state of the art, and where RFID is expected to be taken in the future
- The role of middleware software to route data between the RFID network and the IT systems within an organization
- The use of RFID technology in both commercial and government applications
- The role and value of RFID industry standards and the current regulatory compliance environment
- The issues faced by the public and industry regarding the deployment of RFID technology

An RFID system is composed of three basic components: a tag, a reader, and a host computer.

RFID tags contain tiny semiconductor chips and miniaturized antennas inside some form of packaging. They can be uniquely identified by the reader/host pair and, when applied or fastened to an object or a person, that object or person can be tracked and identified wirelessly. RFID tags come in many forms. For example, some look like paper labels and are applied to boxes and packaging; others are incorporated into the walls of injection molded plastic containers; and still others are built into wristbands and worn by people.

There are many types of RFID tags. Some include miniature batteries that are used to power the tag, and these are referred to as *active* tags. Those that don't include an on-board battery have power "beamed" to them by the reader and are called *passive* tags. In addition, some tags have memories that can be written to and erased, like a computer hard disk, while others have memories that can only be read, like a CD-ROM; these are referred to as "smart" and read-only tags, respectively. The cost and performance of tags can vary widely depending on which of these features are included in their design.

RFID tags can hold many kinds of information about the objects they are attached to, including serial numbers, time stamps, configuration instructions and much more.

RFID readers are composed of an antenna and an electronics module. The antenna is used for communicating with RFID tags wirelessly. The electronics module is most often networked to the host computer through cables and relays messages between the host computer and all the tags within the antenna's read range. The electronics module also performs a number of security functions such as encryption/decryption and user authentication, and another critical function called *anti-collision*, which enables one reader to communicate with hundreds of tags simultaneously.

RFID hosts are the "brains" of an RFID system and most often take the form of a PC or a workstation. (Following this analogy, the readers would constitute the nervous system, while the tags are the objects to be sensed.) Most RFID networks are composed of many tags and many readers. The readers, and consequently the tags, are networked together by means of the central host. The information collected from the tags in an RFID system is processed by the host. The host is also responsible for shuttling data between the RFID network and larger enterprise IT systems, where supply chain management or asset management databases may be operating.

It is believed that RFID technology may someday replace bar codes. While bar code tags and bar code systems are much less expensive than RFID at present, RFID provides many benefits that bar code systems cannot, such as:

- The ability to both read *and* write to tags
- Higher data rates and larger memory sizes
- The ability to function without a direct line of sight between tag and reader
- The ability to communicate with more than one tag simultaneously
- Greater data security (through greater complexity and encryption)
- Greater environmental durability (in the presence of dirt, water, etc.)

The Wal-Mart and DoD mandates are driving the current explosion in the RFID growth. The recent emergence of RFID technology standards, particularly the EPC standard published by EPCglobal, have also encouraged the growth of the industry.

In 2005, Wal-Mart's and DoD's top 100 suppliers began tagging pallets of merchandise. By late 2007, the price of RFID tags, will have dropped to $0.05 it is predicted and RFID will be widespread. In the next 10 years, item-level tagging of merchandise will become commonplace and RFID technology will be ubiquitous, the way television, PC's, and mobile phones already are.

In order to reap the full benefits of RFID, those who implement RFID solutions must find ways to incorporate RFID data into their decision-making processes. Enterprise IT systems are central to those processes. Thus, not unless RFID systems are merged into enterprise IT systems will the companies and organizations that invest in RFID be able to improve business and organizational processes and efficiencies.

This is where middleware comes in. Middleware is the software that connects new RFID hardware with legacy enterprise IT systems. It is responsible for the quality and ultimately the usability of the information produced by RFID systems. It manages the flow of data between the many readers and enterprise applications, such as supply chain management and enterprise resource planning applications, within an organization.

RFID middleware has four main functions:

- Data Collection—Middleware is responsible for the extraction, aggregation, smoothing, and filtering of data from multiple RFID readers throughout an RFID network.
- Data Routing—Middleware facilitates the integration of RFID networks with enterprise systems. It does this by directing data to appropriate enterprise systems within an organization.
- Process Management—Middleware can be used to trigger events based on business rules.
- Device Management—Middleware is also used to monitor and coordinate readers.

The main feature of RFID technology is its ability to identify, locate, track, and monitor people and objects without a clear line of sight between the tag and the reader. Addressing some or all of these functional capabilities ultimately defines the RFID application to be developed in every industry, commerce, and service where data needs to be collected.

In the near-term commercial applications of RFID technology that track supply chain pallets and crates will continue to drive development and growth, however, the Wal-Mart and DoD mandates have also generated interest in the development of other RFID applications outside the commercial retail area, such as RFID-enabled personal security and access control devices.

Security management-related RFID applications enable comprehensive identification, location, tracking, and monitoring of people and objects in all types of environments and facilities.

The applications for RFID technology at present can be categorized as follows:

- Retail and Consumer Packaging—Inventory and supply management chain management, point of sale applications, and pallet and crate tracking
- Transportation and Distribution—Trucking, warehouses, highway toll tags, and fleet management, etc., to monitor access and egress from terminal facilities, transaction recording, and container tracking.
- Industrial and Manufacturing—In a production plant environment, RFID technology is ideally suited for the identification of high-value products moving through a complex assembly process where durable and permanent identification from cradle to grave is essential.
- Security and Access Control—High value asset tracking, building/facility access control, identification card management, counterfeit protection, computer system access and usage control, branded goods replication prevention, baggage handling, and stolen item recovery.

Federal, state, and local governments are taking a larger role in the deployment of RFID technology. DoD is currently one of the leaders in the government's use of RFID technology and is engaged in developing innovative uses of the technology from tracking items within its supply chain to tracking armaments, food, personnel, and clothing to war theaters. Other federal agencies are rapidly following suit with their own RFID projects.

As a technological solution to a complex and far-reaching problem, RFID technology is well suited to improving homeland security. It has many inherent qualities and capabilities that support (1) identity management systems and (2) location determination systems that are fundamental to controlling the U.S. border and protecting transportation systems.

Two of the major initiatives of the border and transportation security strategy that will require extensive use of RFID technology are:

- Creating "smart borders"—At our borders, the DHS could verify and process the entry of people in order to prevent the entrance of contraband, unauthorized aliens, and potential terrorists.
- Increasing the security of international shipping containers—Containers are an indispensable but vulnerable link in the chain of global trade; approximately 90% of the world's cargo moves by container. Each year, nearly 50% of the value of all U.S. imports arrives via 16 million containers. Very few containers coming into the United States are checked.

DHS has initiated the first part its RFID technology program through the U.S.-VISIT initiative, which currently operates at 115 airports and 14 seaports. U.S.-VISIT combines RFID and biometric technologies to verify the identity of foreign visitors with non-immigrant visas.

RFID technology makes immediate economic sense in areas where the cost of failure is great. Homeland security is one area where a high premium can be placed on preventing problems before they occur. Accordingly, for the foreseeable future, developing effective homeland security RFID applications will continue to be a stimulus and driver in RFID technology development.

Wal-Mart and the DoD both specified the use of EPCglobal RFID technology standards in their RFID mandates described in the attached Appendices. Other major retailers, such as Target and Metro AG, the leading retailer in Germany, have also adopted the standards developed by EPCglobal. As a result, the EPCglobal standards appear to be the standards of choice for retailing and supply chain management applications, and it is believed that their standards will have a great influence over the direction the technology and industry ultimately takes.

A number of important implementation issues still need to be addressed before there is widespread adoption of RFID technology. The most important impediments in the development of RFID technology are:

- Resolving consumer privacy issues
- Overcoming the higher costs of developing and deploying RFID technology compared with traditional bar code technology
- Technological immaturity and integration with legacy data management systems
- Need for RFID tag and system robustness
- Lack of application experience, end-user confusion, and scepticism
- Insufficient training and education on RFID applications
- Scope, utilization, and cost of data management tools

In the U.S. consumer-driven economy, personal privacy is protected by a complex and interrelated structural body of legal rights and regulations, consumer protections, and industry and business policy safeguards. To privacy advocates, RFID technology has the potential of weakening these personal privacy protections. According to privacy advocates, RFID technology, if used improperly, jeopardizes consumer privacy, reduces or eliminates purchasing anonymity, and threatens civil liberties.

In comparison to the use of bar codes, RFID technology is still a complex technology in which wide-scale experience is limited. Knowledge and training for the use of RFID technology is relatively low in most organizations. Installation of RFID technology currently lies with smaller companies and vendors that are involved in the initial projects and installations. With time, this will change to participation on a broader scale by mid- and large-size organizations. In order to obtain widespread development of RFID technology it will require the participation, support, knowledge, and data integration expertise of much larger technology development and data management companies.

RFID is here to stay. In the coming years, RFID technology will slowly penetrate many aspects of our lives.

Those companies and government organizations that decide to research and invest in the technology now will not only become the early winners but also derive a benefit from their early knowledge when extending the technology to new applications in the future.

ACKNOWLEDGMENTS

RFID-A Guide to Radio Frequency Identification has been written based on information from a wide variety of authorities who are specialists in their respective fields.

Information in this book has been based in whole or in part on various printed sources or Internet web pages. Direct quotes or selected graphics are used with the permission of the copyright holder.

The author appreciates the efforts by the following individuals to enhance our understanding of radio frequency identification (RFID) technology and products:

Russ Adams, Steve Banker, Raghu Das, Dr. Daniel W. Engles, Rollin Ford, Harris Gardiner, Jeremy Landt, Simon Langford, Tony Seideman, David Williams, and Peter Winer.

The author also appreciates the efforts by the following corporations or organizations for providing information to enhance our understanding of radio frequency identification (RFID) technology and products:

ABI Research, Alanco Technologies Inc., Albertson's, Accenture Corporation, AIM Inc., Applied Business Intelligence, Applied Digital Solutions, Auto-ID Center, Barcodeart, Benetton Clothing Company, Best Buy, Check Point, Coca-Cola, Consumers Against Supermarket Privacy Invasion and Numbering, CVS, Electronic Frontier Foundation, Electronic Privacy Information Center, EPC-globalUS, E-Z Pass Interagency Group, ExxonMobil, General Electric, Gillette, GlaxoSmithKline, Cisco Systems, HD Smith, Hewlett Packard, IDTechEx, IBM, International Standards Organization, Intermec, Impinj, Inc., Johnson &

Johnson, Kraft Foods, LARAN RFID, Los Alamos Scientific National Laboratory, Massachusetts Institute of Technology, Metro, Microsoft Corporation, Motorola, Pfizer, Philips Semiconductor, Port Authority of New York, Proctor & Gamble, Purdue Pharma, RFID Journal, SAP, Sara Lee Foods, SUN, Target, Tesco, Texas Instruments, US Department of Defense, US Department of State, US Department of Justice, US Department of Homeland Security, US Department of Treasury, US Food and Drug Administration, US General Services Administration, US Postal Service, Venture Development Corporation, Wegmans Food Markets, Zebra Technologies Corporation, and other vendors delineated in the RFID Vendor List (See page 157).

We thank Wal-Mart and the Department of Defense for their efforts to Implement RFID tools in the supply chain.

We would also like to thank BuyRFID, formerly known as RFID Wizards Inc. and/or Traxus Technologies, Inc., for permission to reprint graphic material as noted in individual page references throughout this book.

We appreciate the permission to reprint vendor information from the *RFID Journal*.

Also we appreciate the permission to reprint, from the Association for Automatic Identification and Mobility; AIM Inc., their Glossary White Paper Document Version 1.2,2001-08-23, which appears in the Glossary of Terms at the end of the book. Copyright © AIM Inc.; www.aimglobal.org: www.RFID.org.

STAFF ACKNOWLEDGMENTS

The preparation of a book of this type is dependent upon an excellent staff and we have been fortunate in this regard. We appreciate the artwork for this book prepared by Dominic Chiappetta.

This book was prepared as an account of work sponsored by John Wiley & Sons.

Neither the Publisher nor Technology Research Corporation, nor any of its employees, nor any of its contractors, subcontractors, consultants, or their employees, makes any warranty, expressed or implied, or assumes any legal liability or responsibility for the accuracy, completeness, or usefulness or any information, apparatus, product, or process disclosed, or represents that its use would not infringe on privately owned manufacturing rights.

The views, opinions, and conclusions in this book are those of the authors.

Public domain information and those documents abstracted or used in full edited or otherwise used are noted in this acknowledgment or on specific pages or illustrations of this book.

ABOUT THE AUTHORS

V. Daniel Hunt

V. Daniel Hunt is the president of Technology Research Corporation, located in Fairfax Station, Virginia. He is an internationally known management consultant and an emerging technology analyst. Mr. Hunt has 33 years of management and advanced technology experience as part of the professional staffs of Technology Research Corporation, TRW Inc., the Johns Hopkins University/Applied Physics Laboratory, and the Bendix Corporation.

He has served as a senior consultant on projects for the U.S. Department of Defense, the Advanced Research Project Agency, the Department of Homeland Security, the Department of Justice, and for many private firms such as James Martin and Company, Betac Corporation, Lockheed Martin, Northrup Grumman, Hitachi, Pacific Gas and Electric, Electric Power Research Institute, Science Applications International Corporation, Accenture/Arthur Andersen Consulting, and the Dole Foundation.

Mr. Hunt is the author of 20 management and technology professional books. His books include *Process Mapping, Quality in America, Reengineering, Understanding Robotics, Artificial Intelligence and Expert System Sourcebook, Mechatronics*, and the *Gasohol Handbook*. For more information, refer to the web site at http://www.vdanielhunt.com.

Albert B. Puglia

Albert Puglia is an attorney and the senior public safety–privacy issue analyst at Technology Research Corporation.

Since 1997, Mr. Puglia has provided support to the strategic planning and technology management initiatives of the U.S. Department of Justice, U.S.

Department of Homeland Security, and other federal, state, and local law enforcement agencies. He is knowledgeable of current federal, DoD, and state RFID technology initiatives and has worked closely with various public safety agencies in developing and deploying advanced technology.

Mr. Puglia is a former federal law enforcement official, having served in several federal law enforcement agencies, including the U.S. Drug Enforcement Administration and various federal Offices of the Inspector General. His assignments and background in these federal agencies were varied and included operational senior management, organizational assessment, strategic planning, and information systems planning. Mr. Puglia has been recognized for his law enforcement and management leadership and is the recipient of numerous awards and recognition, including the prestigious U.S. Meritorious Service Award.

Mr. Puglia received his B.A. in business administration from Merrimack College, North Andover, Massachusetts, and his M.A. in criminal justice from American University, Washington, D.C.

Mike Puglia

Mike Puglia served as an RFID and advanced wireless engineering technology analyst and writer at Technology Research Corporation. Mr. Puglia has supported Technology Research Corporation technology analysis contracts for various federal agencies, including the U.S. Department of Justice and the U.S. Department of Homeland Security in the area of RFID for public safety applications and emerging technology initiatives.

After graduating from the University of Delaware with a B.S. in electrical engineering and a B.S. in computer engineering, Mr. Puglia worked as an operations engineer at a satellite telecom startup in Annapolis, Maryland. Later he was an RF engineer at Cingular Wireless in San Diego, California, where he designed wireless phone and data networks and developed empirical models for radio wave propagation in urban and suburban environments.

In 2002, Mr. Puglia moved to Asia, where he spent the next two years teaching English in Tokyo and Shanghai and traveling throughout East Asia. During this period, he developed a keen interest in economics, particularly in finance. He is currently completing the Masters of Financial Engineering Program at the University of California at Berkeley. After completing the program, Mr. Puglia will to return to Japan to pursue a career in investment banking.

CHAPTER 1

INTRODUCTION

1.1 WHAT IS RFID?

RFID is an acronym for radio frequency identification, which is a wireless communication technology that is used to uniquely identify tagged objects or people. It has many applications. Some present-day examples include:

- Supply chain crate and pallet tracking applications, such as those being used by Wal-Mart and the Department of Defense (DoD) and their suppliers
- Access control systems, such as keyless entry and employee identification devices
- Point-of-sale applications such as ExxonMobil's Speedpass
- Automatic toll collection systems, such as those increasingly found at the entrances to bridges, tunnels, and turnpikes
- Animal tracking devices, which have long been used in livestock management systems and are increasingly being used on pets
- Vehicle tracking and immobilizers
- Wrist and ankle bands for infant ID and security

The applications don't end there. In the coming years, new RFID applications will benefit a wide range of industries and government agencies in ways that no other technology has ever been able.

RFID-A Guide to Radio Frequency Identification, by V. Daniel Hunt, Albert Puglia, and Mike Puglia
Copyright © 2007 by Technology Research Corporation

1.2 WHAT EXPLAINS THE CURRENT INTEREST IN RFID TECHNOLOGY?

RFID is rapidly becoming a cost-effective technology. This is in large part due to the efforts of Wal-Mart and DoD to incorporate RFID technology into their supply chains.

In 2003, with the aim of enabling pallet-level tracking of inventory, Wal-Mart issued an RFID mandate requiring its top 100 suppliers to begin tagging pallets and cases by January 1, 2005, with Electronic Product Code (EPC) labels. (EPC is the first worldwide RFID technology standard.) DoD quickly followed suit and issued the same mandate to its top 100 suppliers. Since then, Wal-Mart has expanded its mandate by requiring all of its key suppliers to begin tagging cases and pallets. This drive to incorporate RFID technology into their supply chains is motivated by the increased shipping, receiving, and stocking efficiency and the decreased costs of labor, storage, and product loss that pallet-level visibility of inventory can offer.

Wal-Mart and DoD are, respectively, the world's largest retailer and the world's largest supply chain operator. Due to the combined size of their operations, the RFID mandates are spurring growth in the RFID industry and bringing this emerging technology into the mainstream. The mandates are seen to have the following effects:

- To organize the RFID industry under a common technology standard, the lack of which has been a serious barrier to the industry's growth
- To establish a hard schedule for the rollout of RFID technology's largest application to date
- To create an economy of scale for RFID tags, the high price of which has been another serious barrier to the industry's growth

Supply chain and asset management applications are expected to dominate RFID industry growth over the next several years. While presently these applications only account for a small portion of all tag sales, by late 2007, supply chain and asset management applications will account for 70% of all tag sales.[1] As shown in Figure 1-1, the growth in total RFID transponder tags will have grown from 323 million units to 1,621 million units in just five years.

Wal-Mart and DoD alone cannot account for all the current interest in RFID technology, however. Given the following forecasts of industry growth, it becomes clear why RFID has begun to attract the notice of a wide range of industries and government agencies:

---

[1] *RFID White Paper*, Allied Business Intelligence, 2002.

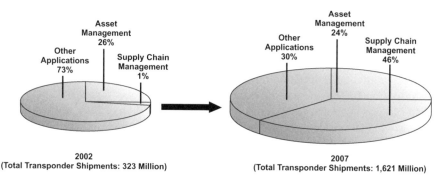

Figure 1-1 Total RFID Transponder Shipments, 2002 vs. 2007. Source: ABI Research.

- In the past 50 years, approximately 1.5 billion RFID tags have been sold worldwide. Sales for 2007 alone are expected to exceed 1 billion and as many as 1 trillion could be delivered by 2015.[2]
- Wal-Mart's top 100 suppliers alone could account for 1 billion tags sold annually.[3]
- Revenues for the RFID industry were expected to hit $7.5 billion by 2006.[4]
- Early adopters of RFID technology were able to lower supply chain costs by 3–5% and simultaneously increase revenue by 2–7% according to a study by AMR Research.[5]
- For the pharmaceutical industry alone, RFID-based solutions are predicted to save more than $8 billion by 2006.[6]
- In the retailing sector, item-level tagging could begin in as early as five years.[7]

In short, the use of RFID technology is expected to grow significantly in the next five years, and it is predicted that someday RFID tags will be as pervasive as bar codes.

---

[2] *RFID Explained*, Raghu Das, IDTechEx, 2004.
[3] *The Strategic Implications of Wal-Mart's RFID Mandate*, David Williams, *Directions Magazine* (www.directionsmag.com), July 2004.
[4] *Radio Frequency Identification (RFID)*, Accenture, 11/16/2001.
[5] *Supply Chain RFID: How It Works and Why It Pays*, Intermec.
[6] *Item-Level Visibility in the Pharmaceutical Supply Chain: A Comparison of HF and UHF RFID Technologies*, Philips Semiconductors *et al*, July 2004.
[7] *Item-Level Visibility in the Pharmaceutical Supply Chain: A Comparison of HF and UHF RFID Technologies*, Philips Semiconductors *et al*, July 2004.

1.3 GOALS OF THIS BOOK

This book provides a broad overview and guide to RFID technology and its application. It is an effort to do the initial "homework" for the reader interested in better understanding RFID tools. It is written to provide an introduction for business leaders, supply chain improvement advocates, and technologists to help them adopt RFID tools for their unique applications, and provide the basic information for better understanding RFID.

The book describes and addresses the following:

- How RFID works, how it's used, and who is using it.
- The history of RFID technology, the current state of the art, and where RFID is expected to be taken in the future.
- The role of middleware software to route data between the RFID network and the information technology (IT) systems within an organization.
- The use of RFID technology in both commercial and government applications.
- The role and value of RFID industry standards and the current regulatory compliance environment.
- The issues faced by the public and industry regarding the wide-scale deployment of RFID technology.

CHAPTER 2

AN OVERVIEW OF RFID TECHNOLOGY

2.1 THE THREE CORE COMPONENTS OF AN RFID SYSTEM

An RFID system uses wireless radio communication technology to uniquely identify tagged objects or people. There are three basic components to an RFID system, as shown in Figure 2-1:

1. A tag (sometimes called a transponder), which is composed of a semi-conductor chip, an antenna, and sometimes a battery
2. An interrogator (sometimes called a reader or a read/write device), which is composed of an antenna, an RF electronics module, and a control electronics module
3. A controller (sometimes called a host), which most often takes the form of a PC or a workstation running database and control (often called middleware) software

The tag and the interrogator communicate information between one another via radio waves. When a tagged object enters the read zone of an interrogator, the interrogator signals the tag to transmit its stored data. Tags can hold many kinds of information about the objects they are attached to, including serial numbers, time stamps, configuration instructions, and much more. Once the interrogator has received the tag's data, that information is relayed back to the controller via a standard network interface, such as an

RFID-A Guide to Radio Frequency Identification, by V. Daniel Hunt, Albert Puglia, and Mike Puglia
Copyright © 2007 by Technology Research Corporation

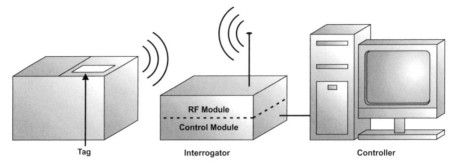

Figure 2-1   The Basic Building Blocks of an RFID System. Source: LARAN RFID.

ethernet LAN or even the internet. The controller can then use that information for a variety of purposes. For instance, the controller could use the data to simply inventory the object in a database, or it could use the information to redirect the object on a conveyor belt system.

An RFID system could consist of many interrogators spread across a warehouse facility or along an assembly line. However, all of these interrogators could be networked to a single controller. Similarly, a single interrogator can communicate with more than one tag simultaneously. In fact, at the present state of technology, simultaneous communication at a rate of 1,000 tags per second is possible, with an accuracy that exceeds 98%.[8] Finally, RFID tags can be attached to virtually anything, from a pallet, to a newborn baby, to a box on a store shelf.

2.2 RFID TAGS

The basic function of an RFID tag is to store data and transmit data to the interrogator. At its most basic, a tag consists of an electronics chip and an antenna (see Figure 2-2) encapsulated in a package to form a usable tag, such as a packing label that might be attached to a box. Generally, the chip contains memory where data may be stored and read from and sometimes written, too, in addition to other important circuitry. Some tags also contain batteries, and this is what differentiates active tags from passive tags.

2.2.1 Active vs. Passive Tags

RFID tags are said to be active if they contain an on-board power source, such as a battery. When the tag needs to transmit data to the interrogator, it uses this source to derive the power for the transmission, much the way a

[8] *Item-Level Visibility in the Pharmaceutical Supply Chain: A Comparison of HF and UHF RFID Technologies*, Philips Semiconductors *et al*, July 2004.

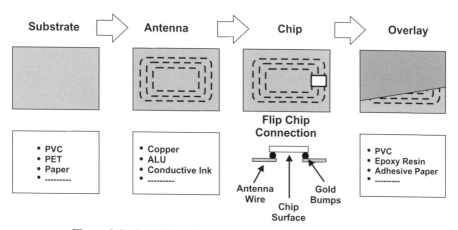

Figure 2-2 RFID Tag Components. Source: LARAN RFID.

cell phone uses a battery. Because of this, active tags can communicate with less powerful interrogators and can transmit information over much longer ranges, up to hundreds of feet. Furthermore, these types of tags typically have larger memories, up to 128 Kbytes.[9] However, they are much larger and more complex than their passive counterparts too, making them more expensive to produce. The batteries in active tags can last from two to seven years.[10]

Passive RFID tags have no on-board power source. Instead, they derive power to transmit data from the signal sent by the interrogator, though much less than if a battery-were on-board. As a result of this, passive tags are typically smaller and less expensive to produce than active tags. However, the effective range of passive tags is much shorter than that of active tags, sometimes under two feet. (Compare a battery-powered megaphone to an old-fashioned plastic cone.) Furthermore, they require more powerful interrogators and have less memory capacity, on the order of a few kilobytes.

Some passive tags do have batteries on-board but do not use these batteries to assist in radio signal transmission. These types of passive tags are called battery-assisted tags and they use the battery only to power on-board electronics. For example, a food producer may apply RFID tags equipped with temperature sensors to pallets in order to monitor the temperature of their product during shipment and storage. Were the temperature of the product to rise above a certain level, that occurrence could be marked on the tag automatically by the sensor. Later, at the time of delivery or sale, the tag could be checked to verify proper shipment or storage. Passive tags equipped with

[9] *RFID Webinar*, www.rfid.zebra.com/RFID_webinar.html, Zebra Technologies.
[10] *Radio Frequency Identification (RFID)*, Accenture, 11/16/2001.

this type of peripheral sensor would need an on-board battery to operate during shipment or storage.

2.2.2   Read-Only vs. Read/Write or "Smart" Tags

Another differentiating factor between tags is memory type. There are roughly two kinds: read-only (RO) and read/write (RW).

RO memory is just that; memory that can be read only. RO tags are similar to bar codes in that they are programmed once, by a product manufacturer for instance, and from thereon cannot be altered, much the way a CD-ROM cannot be altered after it's burned at the factory. These types of tags are usually programmed with a very limited amount of data that is intended to be static, such as serial and part numbers, and are easily integrated into existing bar code systems.

RW tags are often called "smart" tags. Smart tags present the user with much more flexibility than RO tags. They can store large amounts of data and have an addressable memory that is easily changed. Data on an RW tag can be erased and re-written thousands of times, much the same way a floppy disk can be erased and re-written at will. Because of this, the tag can act as a "traveling" database of sorts, in which important dynamic information is carried by the tag, rather than centralized at the controller. The application possibilities for smart tags are seemingly endless. This, in addition to recent advances in smart tag technology that have driven production costs down to under $1 per tag,[11] accounts for much of the present interest in RFID systems.

There are a few variations on these two types of memory that need mentioning. First, there is another memory type called write-once-read-many (WORM). It is similar to RO in that it is intended to be programmed with static information. Drawing on the analogy above, if RO is similar to a CD-ROM, then WORM would be akin to CDRW, in which an end-user, a PC owner for instance, gets one chance only to write in its own information, i.e., burn a blank CD. This type of memory could be used on an assembly line to stamp the manufacturing date or location onto a tag after the production process is complete.

In addition, some tags could contain both RO and RW memory at the same time. For example, an RFID tag attached to a pallet could be marked with a serial number for the pallet in the RO section of the memory, which would remain static for the life of the pallet. The RW section could then be used to indicate the contents of the pallet at any given time, and when a pallet is cleared and reloaded with new merchandise, the RW section of the memory could be re-written to reflect the change.[12]

---

[11] *The Cutting Edge of RFID Technology and Applications for Manufacturing and Distribution*, Susy d'Hunt, Texas Instrument TIRIS.
[12] *Supply Chain RFID: How It Works and Why It Pays*, Intermec.

2.2.3 Tag Form Factors

RFID tags can come in many forms and may not resemble an actual tag at all. Because the chip/antenna assembly in an RFID tag has been made so small, they can now be incorporated into almost any form factor:

- Some of the earliest RFID systems were used in livestock management, and the tags were like little plastic "bullets" attached to the ears of livestock.
- The RFID tags used in automatic toll collection systems are not really tags but plastic cards or key chain type wands.
- In prison management applications, RFID tags are being incorporated into wristbands worn by inmates and guards. Similarly, some FedEx drivers carry RFID wristbands in lieu of a key chain to access their vans through keyless entrance and ignition systems.
- The pharmaceutical industry is incorporating RFID tags into the walls of injection-molded plastic containers, thus blurring the line between tag and packaging.

In short, the form a tag takes is highly dependant upon the application. Some tags need to be made to withstand high heat, moisture, and caustic chemicals, and so are encased in protective materials. Others are made to be cheap and disposable, such as "smart" labels. A "smart" label is just one form of a "smart" tag, in which an RFID tag is incorporated into a paper packing label. While there are many applications in which RFID tags are anything but, the overall trend in the industry is towards this small, flat label that can be applied quickly and cheaply to a box or pallet.

2.3 RFID INTERROGATORS

An RFID interrogator acts as a bridge between the RFID tag and the controller and has just a few basic functions.

- Read the data contents of an RFID tag
- Write data to the tag (in the case of smart tags)
- Relay data to and from the controller
- Power the tag (in the case of passive tags)

RFID interrogators are essentially small computers. They are also composed of roughly three parts: an antenna, an RF electronics module, which is responsible for communicating with the RFID tag, and a controller electronics module, which is responsible for communicating with the controller.

   In addition to performing the four basic functions above, more complex RFID interrogators are able to perform three more critical functions:

- implementing anti-collision measures to ensure simultaneous RW communication with many tags,
- authenticating tags to prevent fraud or unauthorized access to the system,
- data encryption to protect the integrity of data.

2.3.1   Multiple RW and Anticollision

Anticollision algorithms are implemented to enable an interrogator to communicate with many tags at once. Imagine that an interrogator, not knowing how many RFID tags might be in its read zone or even if there are any tags in its read zone, issues a general command for tags to transmit their data. Imagine that there happen to be a few hundred tags in the read zone and they all attempt to reply at once. Obviously a plan has to be made for this contingency. In RFID it is called anticollision.

There are three types of anticollision techniques: spatial, frequency, and time domain. All three are used to establish either a pecking order or a measure of randomness in the system, in order to prevent the above problem from occurring, or at least making the occurrence statistically unlikely.

2.3.2   Authentication

High-security systems also require the interrogator to authenticate system users. Point of sale systems, for example, in which money is exchanged and accounts are debited, would be prone to fraud if measures were not taken. In this very high-security example, the authentication procedure would probably be two-tiered, with part of the process occurring at the controller and part of the process occurring at the interrogator.

There are basically two types of authentication. They are called *mutual symmetrical* and *derived keys*.[13] In both of these systems, an RFID tag provides a key code to the interrogator, which is then plugged into an algorithm, or a "lock," to determine if the key fits and if the tag is authorized to access the system.

2.3.3   Data Encryption/Decryption

Data encryption is another security measure that must be taken to prevent external attacks to the system. In the POS example, imagine that a third party were to intercept a user's key. That information could then be used to make fraudulent purchases, just as in a credit card scam. In order to protect the integrity of data transmitted wirelessly, and to prevent interception by a third party, encryption is used. The interrogator implements encryption and

---

[13] *Radio Frequency Identification (RFID)*, Accenture, 11/16/2001.

decryption to do this. Encryption is also central to countering industrial espionage, industrial sabotage, and counterfeiting.

2.3.4 Interrogator Placement and Form Factors

RFID systems do not require line of sight between tags and readers the way that bar code systems do. As a result of this, system designers have much more freedom when deciding where to place interrogators. Fixed-position interrogators can be mounted in dock doors, along conveyor belts, and in doorways to track the movement of objects through any facility. Some warehousing applications even hang interrogator antennae from the ceiling, along the aisles of shelves, to track the movement of forklifts and inventory.

Portable readers can be mounted in forklifts, trucks, and other material-handling equipment to track pallets and other items in transit. There are even smaller, hand-held portable interrogator devices that enable users to go to remote locations where it's not feasible to install fixed-position interrogators. Often these portable devices are connected to a PC or laptop, either wirelessly or with a cable. These PC's or laptops are in turn networked to the controller, again, either wirelessly or with a cable.[14]

2.4 RFID CONTROLLERS

RFID controllers are the "brains" of any RFID system. They are used to network multiple RFID interrogators together and to centrally process information. The controller in any network is most often a PC or a workstation running database or application software, or a network of these machines. The controller could use information gathered in the field by the interrogators to:

- Keep inventory and alert suppliers when new inventory is needed, such as in a retailing application
- Track the movement of objects throughout a system, and possibly even redirect them, such as on a conveyor belt in a manufacturing application
- Verify identity and grant authorization, such as in keyless entry systems
- Debit an account, such as in Point of Sale (POS) applications

2.5 FREQUENCY

A key consideration for RFID is the frequency of operation. Just as television can be broadcast in a VHF or a UHF band, so too can RFID systems use different bands for communication as shown in Figure 2-3.

---

[14] *Supply Chain RFID: How It Works and Why It Pays*, Intermec.

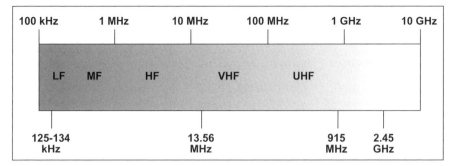

Figure 2-3 Radio Frequency Spectrum. Source: Texas Instruments.

In RFID there are both low frequency and high radio frequency bands in use, as shown in the following list:

Low Frequency RFID Bands

- Low frequency (LF): 125–134 KHz
- High frequency (HF): 13.56 MHZ

High Frequency RFID Bands

- Ultra-high frequency (UHF): 860–960 MHZ
- Microwave: 2.5 GHz and above

The choice of frequency affects several characteristics of any RFID system, as discussed below.

2.5.1 Read Range

In the lower frequency bands, the read ranges of passive tags are no more than a couple feet, due primarily to poor antenna gain. (At low frequencies, electromagnetic wavelengths are very high, on the order of several miles sometimes, and much longer than the dimensions of the antennas integrated into RFID tags. Antenna gain is directly proportional to antenna size relative to wavelength. Hence, antenna gain at these frequencies is very low.) At higher frequencies, the read range typically increases, especially where active tags are used. However, because the high frequency bands pose some health concerns to humans, most regulating bodies, such as the FCC, have posed power limits on UHF and microwave systems and this has reduced the read range of these high frequency systems to 10 to 30 feet on average in the case of passive tags.[15]

---

[15] *Supply Chain RFID: How It Works and Why It Pays*, Intermec.

2.5.2   Passive Tags vs. Active Tags

For historical reasons, passive tags are typically operated in the LF and HF bands, whereas active tags are typically used in the UHF and microwave bands. The first RFID systems used the HF and LF band with passive tags because it was cost prohibitive at the time to do otherwise. Today, however, that is quickly changing. Recent advances in technology have made it feasible to use both active tags and the higher frequency bands and this has been the trend in the industry.

2.5.3   Interference from Other Radio Systems

RFID systems are prone to interference from other radio systems. RFID systems operating in the LF band are particularly vulnerable, due to the fact that LF frequencies do not experience much path loss, or attenuate very little over short distances, in comparison to the higher frequencies. This means that the radio signals of other communication systems operating at nearly the same LF frequency will have high field strengths at the antenna of an RFID interrogator, which can translate into interference. At the other end of the spectrum, microwave systems are the least susceptible to interference, as path loss in the microwave band is much higher than for the lower frequencies, and generally a line of sight is required in order for microwave radiators to interfere.

2.5.4   Liquids and Metals

The performance of RFID systems will be adversely affected by water or wet surfaces. HF signals, due to their relatively long wavelengths, are better able to penetrate water than UHF and microwave signals. Signals in the high frequency bands are more likely to be absorbed in liquid. As a result, HF tags are a better choice for tagging liquid-bearing containers.[16]

Metal is an electromagnetic reflector and radio signals cannot penetrate it. As a result, metal will not only obstruct communication if placed between a tag and an interrogator, but just the near presence of metal can have adverse affects on the operation of a system; when metal is placed near any antenna the characteristics of that antenna are changed and a deleterious effect called de-tuning can occur.

The high frequency bands are affected by metal more so than the lower frequency bands. In order to tag objects made of metal, liquid bearing containers, or materials with high dielectric permittivity, special precautions have to be taken, which ultimately drives up costs.

---

[16] *Item-Level Visibility in the Pharmaceutical Supply Chain: A Comparison of HF and UHF RFID Technologies*, Philips Semiconductors *et al*, July 2004.

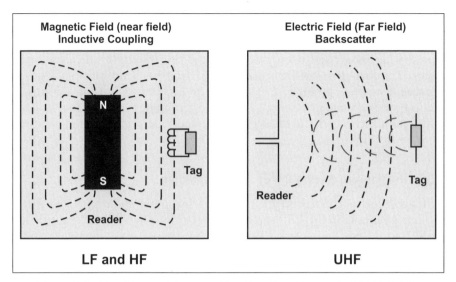

Figure 2-4 Two Types of Antenna/Tag Coupling. Source: LARAN RFID.

2.5.5 Data Rate

RFID systems operating in the LF band have relatively low data rates, on the order of Kbits/s. Data rates increase with frequency of operation, reaching the Mbit/s range at microwave frequencies.[17]

2.5.6 Antenna Size and Type

Due to the long wavelengths of low frequency radio signals, the antennas of LF and HF systems have to be made much larger than UHF and microwave antennas in order to achieve comparable signal gain. This conflicts with the goal of making RFID tags small and cheap, however. Most system designers forsake antenna gain in the name of controlling costs, which ultimately results in a low read range for LF and HF systems. There is a lower limit to how small LF and HF antennas can be made though and as a result, LF and HF tags are typically larger than UHF and Microwave tags.[18] Figure 2-4 shows the two types of RFID antenna/tag coupling concepts.

Frequency of operation will also dictate the type of antenna used in an RF system. At LF and HF, inductive coupling and inductive antennas are used, which are usually loop-type antennas. At UHF and microwave frequencies, capacitive coupling is used and the antennas are of the dipole type.

---

[17] *Radio Frequency Identification (RFID)*, Accenture, 11/16/2001.
[18] *Radio Frequency Identification (RFID)*, Accenture, 11/16/2001.

2.5.7 Antenna Nulls and Orientation Problems

Inductive antennas, such as those used at the LF and HF, operate by "flooding" a read zone with RF radiation. In addition to the long wavelengths of LF and HF, this works to inundate an interrogator's read zone with a uniform signal that will not differ in strength from one end to the other. Dipole antennas on the other hand, such as those used at the UHF and microwave frequencies, operate by spot beaming signals from transmitter to receiver. This, in addition to the relatively short wavelengths of high frequency UHF and microwave signals, gives rise to small ripples in a UHF or microwave interrogator's read zone, so that signal strength will not be uniform from one end of a read zone to the other and will even diminish to zero at some points, creating "nulls," or invisible spots. RFID tags positioned in these null spots are rendered effectively invisible to an RF interrogator, which can obviously cause problems in UHF and microwave systems.

Null spots can also occur from the detuning of tags, which occurs when two tags are placed in close proximity to one another or in close proximity to liquids, metals, and other materials with a high dielectric permittivity.

UHF and microwave systems are more sensitive to differences in antenna orientation as shown in Figure 2-5. Inductive antennas have little directional gain, meaning signals strengths at a given distance are the same above, below, in front or behind the antenna, dipole antennas have a more highly directive gain and significant differences in field strength at a given distance will exist between points in front of the dipole and above it. For UHF and microwave tags oriented top-up to the interrogator (imagine a box on its side passing through a dock door interrogator), signal strengths might not be high enough to enable communication.

All of these phenomena require that UHF and microwave RFID systems implement a more complex form of modulation called frequency hopping to overcome their shortcomings.

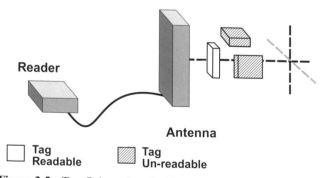

Figure 2-5 Tag Orientation Problems. Source: LARAN RFID.

2.5.8   Size and Price of RFID Tags

Early RFID systems used primarily the LF band, due to the fact that LF tags are the easiest to manufacture. They have many drawbacks, however, such as a large size, as mentioned previously, which translates into a higher price at volume. The HF band is currently the most prevalent worldwide, because HF tags are typically less expensive to produce than LF tags. The UHF band represents the present state of the art. Recent advances in chip technology have brought prices for UHF tags down to the point that they are competitive with HF tags. Microwave RFID tags are similar to UHF tags in that they can be made smaller and ultimately cheaper. Table 2-1 illustrates the RFID system characteristics at various frequencies.

2.6   AUTOMATIC IDENTIFICATION AND DATA CAPTURE (AIDC) SYSTEMS

RFID smart labels trace their origins all the way back to traditional paper tagging. Paper tagging systems, which leverage technology very little, began being replaced in industry in the 1970s by a broad class of technologies called Automatic Identification and Data Capture (AIDC) technologies. RFID is just one part of this family of technologies. Other members include the familiar bar code, as well as optical character recognition (OCR) and infrared identification technologies.[19] RFID could be called the rising star in this family, in that it seems poised to offer many benefits not yet offered by any other technology.

2.6.1   Optical Character Recognition (OCR)

OCR systems are able to optically scan text on a printed page and convert the image into a text file that can be manipulated by a computer, such as an ASCII file or an MS Word document. (A computer is not able to see a pure image file as other than a collection of white and black dots on a page. An ASCII file or an MS Word document, in contrast, is viewed by the computer as a collection of letters on a page. As a result, ASCII files and MS Word documents can be edited, searched for text, spell-checked, data-based, etc., while image files can't.) Using this type of technology, an entire book could be scanned with a desktop scanner and converted into a text document. Similarly, in a retailing application, a paper price tag *could* be read this way at checkout, and the information in the text file produced could be used to write up a sales slip, inventory the item or charge a credit card account. This, however, would not be a very efficient way of doing things. While there are applications in which OCR technology is superior to RFID, such as in the legal profession, where searches that once took days have been whittled down to a few minutes, in

[19] *Radio Frequency Identification (RFID)*, Accenture, 11/16/2001.

TABLE 2-1   RFID System Characteristics at Various Frequencies

Frequency Band	LF 125 KHz	HF 13.56 MHZ	UHF 860–960 MHZ	Microwave 2.5 GHz and Up
Read Range (Passive Tags)	<2 Feet	<3 Feet	<10–30 Feet	–10 Feet
Tag Power Source	Generally passive	Generally passive	*Generally active but* Passive Also	*Generally active but* Passive Also
Tag Cost	Relatively expensive	*Expensive*, but less So Than LF	Potential to Be very cheap	Potential to Be very cheap
Typical Applications	Keyless entry, animal tracking, vehicle immobilizers, POS	"Smart" cards, item-level track such as baggage handling, libraries	Pallet tracking, electronic toll collection, baggage handling	Electronic toll collection
Data Rate	Slower			Faster
Performance Near Metal or Liquids	Better			Worse
Passive Tag Size	Larger			Smaller

Source: ABI Research.

supply chain and asset management applications RFID is still the AIDC technology of choice.

2.6.2 Infrared Identification

Infrared identification technology is very similar to RFID technology, the main difference being the frequency of operation. In the electromagnetic spectrum, the infrared frequencies are far higher than even the highest microwave frequencies used in RFID. At infrared frequencies path losses are very high and infrared signals are not able to penetrate solid objects very well, such as boxes, to read tags. As a result, infrared identification is used more often in imaging applications, such as night vision and motion detection.

2.6.3 Bar Codes

A bar code is a series of vertical, alternating black and white stripes of varying widths that form a machine-readable code. Bar coding is an optical electronic technology, in which laser light is reflected off a bar code symbol and read by a scanner.

The ubiquitous Universal Product Code (UPC) symbol is the form of bar code familiar to most people. Research in bar coding was begun long before the emergence of the UPC standard, however. In 1952, two researchers at IBM were awarded the first patent for automatic identification technology. They continued to develop the early bar code technology through the 1950s and were joined by others who saw the potential for it. In the 1960s, the first commercial systems emerged, aimed primarily at the rail freight and product distribution industries. Then, in the early 1970s, a consortium of U.S. grocery stores convened an ad hoc committee to evaluate bar coding technology, with the aim of deploying it in supermarkets across the country as a means of driving down labor costs, improving checkout speed and tracking sales and inventory. In 1973 the UPC, shown in Figure 2-6, was born of this effort and became a major driver in the deployment of bar code technology. Figure 2-6, 2-7 and 2-8 show different UPCs.

Growth in grocery store bar coding was slow throughout the 1970s. This was not due to the lack of interest on the part of grocery stores, but rather because product manufacturers were slow to include the symbols on their packaging. It was deemed that a minimum of 85% of all supermarket products would need to include the label before the systems could pay for themselves.[20] In 1978, this mark was reached and sales in bar code scanning systems began to take off. Then, in 1981, DoD initiated the LOGMARS program, which required that all products sold to the military be marked with Code 39 symbols, as shown in Figure 2-8 (another bar code standard, different than UPC).[21]

---

[20] *The History of Bar Codes* (www.basics.ie/History.htm), Tony Seideman.
[21] *Bar Code History Page* (www.adams1.com/pub/russadam/history.html), Russ Adams.

Figure 2-6   UPC A Symbol. Source: www.barcodeart.com.

Figure 2-7   UPC E Symbol. Source: www.barcodeart.com.

Figure 2-8   Code 39 Bar Code Symbol. Source: www.barcodeart.com.

These last two events triggered a revolution in supply chain management. In 1978, for instance, only 1% of grocery stores had bar-coding scanners. By 1981, that number had risen to 10% and by 1984 it was 33%. Today, bar coding technology is used in more than 60% of all grocery stores nationwide.[22] World-wide there are now more than nine bar code standards in use.

2.6.4 RFID "Smart" Labels

RFID smart labels are considered to be the next generation bar code. Just as the bar code sparked a revolution in supply chain and asset management in the early 1980s, smart labels seem poised to do the same in the coming years. As mentioned previously, a smart label is just a RW transponder that has been incorporated into a printed packing label. Like bar codes, these labels are meant to be easily applied, unobtrusive, quick to read, cheap, and disposable.

---

[22] *The History of Bar Codes* (www.basics.ie/History.htm), Tony Seideman.

Some RFID technology manufacturers have made implementing RFID technology as simple as printing out a document on a PC. There are several that now offer smart label printer solutions, which both print out adhesive smart label tags and write data to tag memory. There are even some hybrid bar code/smart tag solutions that both print a UPC bar code symbol on an adhesive smart tag and write data to tag memory simultaneously, in order to assist customers in migrating between the technologies.

There are many measures by which RFID smart labels do not yet stack up to bar codes, such as price, technological maturity, and ease of implementation. However, the benefits that smart labels offer over bar coding systems are beginning to outweigh the shortcomings and the costs of implementing smart labels solutions, making smart labels a cost-effective technology.

2.7 "SMART" TAGS VS. BAR CODES

In bar coding, laser light is used as the data carrier. In contrast, smart labels and RFID in general uses radio waves to carry information. Bar coding is therefore referred to as an optical technology and RFID is called a radio frequency or RF technology. This has several implications for AIDC. Below is a detailed comparison of RFID to bar codes.

2.7.1 Memory Size/Data Storage

Bar codes can only hold a limited amount of data. The smallest tags, in terms of data storage, are UPC E symbols, which hold only eight numeric characters; just a few bytes. At the opposite end of the spectrum, the Data Matrix bar code standard permits the storage of 2000 ASCII characters, on a two-dimensional tag, as shown in Figure 2-9, though these are rarely used.

Figure 2-9 Data Matrix Bar Code Symbol. Source: www.barcodeart.com.

RFID tags are capable of holding far more information. Though RFID tags can be made with smaller memories to hold only a few bytes, the current state of technology puts the upper limit at 128 K bytes, orders of magnitude larger than most bar code symbols.

2.7.2 Read/Write

Bar codes are not able to be modified once they are printed, therefore bar coding is a RO technology. In contrast, RW RFID tags, such as smart tags, have an addressable, writeable memory that can be modified thousands of times over the life of the tag and this is, in part, what makes RFID technology so powerful.

2.7.3 Non-Line-of-Sight

Another advantage of RFID technology over bar codes is that RFID systems do not require a line-of-sight between a tag and interrogator to work properly. Because radio waves are able to propagate through many solid materials, RFID tags buried deep within the contents of a pallet are really no less visible to interrogators than exposed tags with a direct line of sight. In addition, tags embedded inside objects, and not just applied to packaging, can also be read with no problems. Bar codes, on the other hand require a direct line of sight with the scanner in order to work properly. This means that bar codes must be placed on the outside of packaging and objects must be removed from pallets in order to be read. In supply chain management applications, in which large quantities of materials are on the move all the time, this gives RFID a great advantage over bar codes.

2.7.4 Read Range

The read range of bar codes can be quite long. Bar code scanners can be made to scan tags up to several yards away, though only under certain conditions and not without a direct line of sight. Typically read ranges are just a few inches, however. The read ranges of RFID tags vary widely, depending on frequency of operation, antenna size and whether the tag is active or passive. Typically though, read ranges of RFID tags run from a few inches to a couple of yards.

2.7.5 Multiple RW and Anticollision

Unlike other AIDC technologies, in which items must be physically separated and read individually, RFID systems can read multiple tags simultaneously. Whereas a pallet of bar-coded items would need to be unpacked and scanned individually in order to be inventoried, in RFID systems the entire contents of a pallet could be inventoried at once as it passes an interrogator. RFID is the only AIDC technology that is capable of this and the advantages it gives

RFID over bar coding and other systems in supply chain applications cannot be understated.

2.7.6 Access Security

Bar code data is not very secure. Because bar codes require a line-of-sight and are therefore placed very visibly on the outside of packaging, anyone with a standard bar code scanner or even a camera can intercept and record the data. RFID systems offer a much higher level of security. As mentioned previously, RFID systems present the user with the ability to prevent third-party interception, to restrict unauthorized access to the system, and to encrypt sensitive data.

2.7.7 Difficult to Replicate

Because RFID tags and electronics are so much more complex than bar codes and bar code electronics, RFID systems are much more difficult to build or replicate. This makes it difficult for would-be cheats to access or alter tag data. (For instance, somebody who tries to change the price of an item on a store shelf with a homemade interrogator).

2.7.8 Environmental Susceptibility/Durability

RFID technology is better able to cope with harsh and dirty environments, such as those found in warehouses and supply chain facilities, than bar codes. Bar codes can not be read if they become covered in dirt, dust, or grease or are torn or dented. Intense light can also interfere with bar code scanners and render them unable to read bar code tags. RFID technology is relatively immune to these problems.

2.7.9 Read Reliability

In supply chain applications, first-pass read accuracy is important to maintaining a high level of efficiency. Damaged bar codes often have to be scanned through a system two times or manually read. The anticollision and multiple RW features of RFID eliminate the need to scan misread items multiple times.

2.7.10 Price

The largest barrier to RFID growth is tag cost. Whereas bar codes typically cost under $0.01,[23] the current cost of a passive RFID tag with a read range of a few centimeters is much higher. Reports vary widely, but most put the cost somewhere in the tens of cents range. Production costs for RFID tags can be broken down as follows[24]:

---

[23] *RFID Explained*, Raghu Das, IDTechEx, 2004.
[24] *Radio Frequency Identification (RFID)*, Accenture, 11/16/2001.

TABLE 2-2   Comparison of Bar Code vs. RFID System Characteristics

System	Bar Code	RFID
Data Transmission ↙	Optical	Electromagnetic
Memory/Data Size	Up to 100 bytes	Up to 128 kbytes
Tag Writable	No	Possible
Position of Scan/Reader	Line-of-sight	Non-line-of-sight possible
Read Range	Up to several meters (line-of-sight)	Centimeters to meters (system dependent)
Access Security	Low	High
Environmental Susceptibility	Dirt	Low
Anticollision	Not possible	Possible
Price	<$0.01	$0.10 to $1.00 (passive tags)

Source: Accenture.

- Silicon die production (7–12 cents)
- Die placement on printed circuit board (10 cents)
- Antenna/adhesive packaging (5 cents)
- Shipping and handling expenses

It's difficult to see tag prices falling below production costs. From the above cost analysis, the lower limit on tag prices at present could be assumed to be around $0.30. More complex RFID tags can cost tens of dollars.

RFID technology is predicted to grow tremendously in the coming years and, as a result, an economy of scale is sure to be realized. Some predict the cost of RFID tags for tagging cartons and pallets will fall to $0.05 per tag during 2007, with annual sales of 10 billion tags.[25]

Of course it is not sufficient to compare the costs of bar codes and RFID tags without taking benefits offered into account. There are many applications in which the higher costs of RFID tags more than pay for themselves. For instance, when tracking high value items (such as pharmaceuticals) or reusable containers, a costly RFID tag can still be cost effective. The added efficiency offered by RFID systems can also justify their relatively higher costs. Table 2-2 compares bar code versus RFID system characteristics.

2.8   RFID TECHNOLOGY IN SUPPLY CHAIN MANAGEMENT[26]

The ability to uniquely identify items throughout a supply chain, without line-of-sight, can have many benefits:

[25] *RFID Explained*, Raghu Das, IDTechEx, 2004.
[26] *Item-Level Visibility in the Pharmaceutical Supply Chain: A Comparison of HF and UHF RFID Technologies*, Philips Semiconductors *et al*, July 2004.

2.8.1 Visibility and Efficiency

RFID provides 100% visibility of inventory in a supply chain, regardless of its location. Goods can be moved more easily and more quickly within the supply chain as a result. In addition, productivity in shipping and receiving can be improved, touch labor reduced, shipping accuracy increased, and product availability at retail locations can be expanded through the use of RFID technology.

2.8.2 Accountability and Brand Protection

RFID provides accountability at every point in a supply chain. Inventory losses and write-offs due to shrinkage can be dramatically reduced by having a more accountable supply chain. The ability to track items throughout a supply chain can help in preventing these losses, as well as "gray market" distribution (diversion to unauthorized retail channels), which can cost hundreds of millions of dollars every year.

2.8.3 Product Safety and Recalls

RFID can provide the ability to more closely track lot and expiration dates of merchandise, thereby improving expiration management. In addition, the ability to uniquely identify manufactured items can "reduce the time spent identifying products targeted for recall as well as reducing the likelihood of a mass market recall of branded products."[27]

---

[27] *Item-Level Visibility in the Pharmaceutical Supply Chain: A Comparison of HF and UHF RFID Technologies*, Philips Semiconductors *et al*, July 2004.

CHAPTER 3

HISTORY AND EVOLUTION OF RFID TECHNOLOGY

It is difficult to trace the history of RFID technology back to a well-defined starting point; there is no clear progression of RFID developments over time that ultimately arrives at the present state of matters. Rather, the history of RFID technology is intertwined with that of the many other communications technologies developed throughout the 20th century. These technologies include computers, information technology, mobile phones, wireless LANs, satellite communications, GPS, etc. With RFID just beginning to emerge as a separate technology, it is only in hindsight that we know many of the developments made in these other technologies to have also been developments in RFID technology research, development, and deployment.

3.1 THE CONVERGENCE OF THREE TECHNOLOGIES

Research and advances in the following three areas have given rise to commercially viable RFID:

- Radio Frequency Electronics—Research in this field, as applied to RFID, was begun during WWII and continued through the 1970s. The antenna systems and RF electronics employed by RFID interrogators and tags have been made possible because of radio frequency electronic research and development.

RFID-A Guide to Radio Frequency Identification, by V. Daniel Hunt, Albert Puglia, and Mike Puglia
Copyright © 2007 by Technology Research Corporation

- Information Technology—Research in this field began in the mid-1970s and continued through the mid-1990s roughly. The host computer and the interrogator both employ this technology. The networking of RFID interrogators and the networking of RFID systems (the EPC Network for example) has also been made possible by research in this area.
- Materials Science—Breakthroughs in materials science technology in the 1990s finally made RFID tags cheap to manufacture and, at present, $0.05 tags are on the horizon. Overcoming this cost barrier has gone a long way to making RFID technology commercially viable.

3.2 MILESTONES IN RFID AND THE SPEED OF ADOPTION[28]

In order to better define the development of RFID technology the following time-based development summaries are shown below.

3.2.1 Pre-1940s

The last half of the 19th century saw many advances in our understanding of electromagnetic energy. By the turn of that century, the works of Faraday, Maxwell, Hertz, and others had yielded a complete set of laws describing its nature. Beginning in 1896, Marconi, Alexanderson, Baird, Watson, and many others sought to apply these laws in radio communications and radar. The work done in this era form the building blocks upon which many technologies have been built, including RFID.

3.2.2 1940s—WWII

WWII brought about many advancements in radio frequency communications and radar. Following the war, scientists and engineers continued their research in these areas and increasingly sought civilian uses for it. In October of 1948, Harry Stockman published a paper in the *Proceedings of the IRE* titled "Communications by Means of Reflected Power," which in hindsight may be the closest thing to the birth of RFID technology.

3.2.3 1950s—Early Exploration of RFID Technology

During the 1950s, many of the technologies related to RFID were explored by researchers. A couple of important papers were published, notably F.L. Vernon's "Applications of the Microwave Homodyne" and D.B. Harris's "Radio Transmission Systems with Modulatable Passive Responders." The U.S. military began to implement an early form of aircraft RFID technology called Identification, Friend or Foe, or IFF.

---

[28]*Shrouds of Time: The History of RFID*, Jeremy Landt, *et al*, AIM, October 2001.

3.2.4 1960s—Development of RFID Theory and Early Field Trials

The 1960s were a prelude to an RFID explosion that would come later, in the 1970s. R.F. Harrington did a great deal of research in the field of electromagnetic theory as it applied to RFID, as described in "Field Measurements Using Active Scatterers" and "Theory of Loaded Scatterers."

RFID inventors and inventions began to emerge also. Examples include Robert Richardson's "Remotely Activated Radio Frequency Powered Devices," Otto Rittenback's "Communication by Radar Beams," J.H. Vogelman's "Passive Data Transmission Techniques Utilizing Radar Beams," and J.P. Vinding's "Interrogator-Responder Identification System."

Some commercial activities began in the late 1960s, too. Sensormatic and Checkpoint were founded to develop electronic article surveillance (EAS) equipment for anti-theft and security applications. (Anti-theft gates placed at the doors to department stores for instance.) Their systems were simple, 1-bit systems, meaning they could only detect the presence of RFID tags, rather than identify them. EAS later became the first widespread commercial use of RFID.

3.2.5 1970s—An RFID Explosion and Early-Adopter Applications

The 1970s witnessed a great deal of growth in RFID technology. Companies, academic institutions, and government laboratories became increasingly involved in RFID.

Notable advances were made in research. In 1975, Los Alamos Scientific Laboratory released a great deal of its RFID research to the public in a paper titled "Short-Range Radio-telemetry for Electronic Identification Using Modulated Backscatter," written by Alfred Koelle, Steven Depp, and Robert Freyman.

Large companies such as Raytheon, RCA, and Fairchild began to develop electronic identification system technology, too. By 1978, a passive microwave transponder had been accomplished.

Several government agencies began to show interest in the technology also. The Port Authority of New York and New Jersey experimented with transportation applications developed by GE, Westinghouse, Philips, and Glenayre, though the technology was not adopted. The U.S. Federal Highway Administration convened a conference to explore the use of electronic identification technology in vehicles and transportation applications as well.

Numerous small companies focused on RFID technology began to emerge in the late 1970s. By the end of the decade, much of the research in RF electronics and electromagnetics, as applied to RFID, was complete and research in computers and information technology, crucial to the development of RFID hosts, networks and interrogators, had begun, as evidenced by the birth of the PC and the ARPANET, predecessor to the internet.

3.2.6   1980s—Commercialization

The 1980s brought about the first widespread commercial RFID systems. The systems were simple ones. Examples include livestock management, keyless entry, and personnel access systems. The Association of American Railroads and the Container Handling Cooperative Program became active in RFID initiatives, with the aim of RFID-enabling railroad cars. Transportation applications emerged late in the decade. The world's first toll application was implemented in Norway in 1987, followed by Dallas in 1989. The Port Authority of New York and New Jersey implemented a commercial project for buses passing through the Lincoln Tunnel.

All of the RFID systems implemented in the 1980s were proprietary systems. There was no interoperability between systems and little competition in the RFID industry as a result, which kept costs high and impeded industry growth.

3.2.7   1990s—RFID Enters the Mainstream

The 1990s were significant in that RFID finally began to enter the mainstream of business and technology. By the middle of the decade, RFID toll systems could operate at highway speeds, meaning drivers could pass through toll points unimpeded by plazas or barriers. In addition, it became possible to enforce tolls with video cameras. Deployment of RFID toll systems became widespread in the United States as a result. Regional toll agencies took the technology one step further and began to integrate their RFID systems too, enabling drivers to pay multiple tolls through the same account. Examples include the E-Z Pass Interagency Group, located in the northeastern United States, a project in the Houston area, a project linking toll systems in Kansas and Oklahoma, as well as a project in Georgia.

Texas Instruments began its TIRIS system in the 1990s also. This system developed new RFID applications for dispensing fuel, such as ExxonMobil's Speedpass, as well as ski pass systems and vehicle access systems. In fact, many companies in the United States and Europe became involved in RFID during the 1990s; examples include Philips, Mikron, Alcatel, and Bosch.

Research in information technology was well developed by the early 1990s, as evidenced by the proliferation of PC's and internet. This left the RFID industry with only the problem of expensive tags to overcome, in order to realize commercially viable systems. Advances in materials technology during the 1990s, many of them related to the work of semiconductor chip makers such as IBM, Intel, AMD, and Motorola, finally put cost-effective tags on the horizon. Investment capital began to flow towards RFID and many venture capital projects got underway as a result. Large-scale "smart label" tests had begun by the end of the decade.

Until the 1990s the RFID systems on the market were proprietary systems. Many in the industry recognized this as a barrier to growth and an effort to standardize the technology began. Several standards organizations got to work on publishing guidelines, including the European Conference of Postal and Telecommunications Administrations (CEPT) and the International Organization of Standards (ISO). The Auto-ID Center at M.I.T. was established in 1999 for that purpose also. Currently, all of these organizations are working on standards for RFID technology, particularly supply chain and asset management applications.

3.2.8 2000s—RFID Deployment

By the early 2000s it had become clear that $0.05 tags would be possible and that RFID technology could someday replace bar code systems. The implications this had for the product distribution and retail industries, and the dollar figures involved, garnered a lot of attention for the industry. The year 2003 in particular was an eventful one for RFID. Both Wal-Mart and the DoD, the world's largest retailer and the world's largest supply chain, respectively, issued RFID mandates requiring suppliers to begin employing RFID technology by 2005. The combined size of their operations constitute an enormous market for RFID. Other retailers and many manufacturers, such as Target, Proctor & Gamble, and Gillette, have followed suit.

Furthermore, in 2003, the Auto-ID Center was merged into EPCglobal, a joint venture between the Uniform Product Code Council, makers of the UPC bar code symbol, and EAN. EPC's technology has been adopted by both Wal-Mart and DoD and the RFID industry. It appears that RFID finally has a common platform from which to move forward. The standards developed by EPC were adopted by the ISO in 2006, giving the RFID industry a single source to go to for guidance. The convergence of all standards to one will serve to increase competition amongst players in the industry, lower the costs of RFID and quicken the deployment of RFID technology. (Standards will be discussed later in greater detail.)

As of 2007, it is obvious that numerous applications for RFID across a number of industries will soon emerge. In the coming years, RFID technology will grow further and further into the mainstream and become another part of everyday life, just as television, PC's, and mobile phones already have.

3.3 RFID IN THE FUTURE

With big companies such as Wal-Mart, Proctor & Gamble, Target and Gillette investing heavily in the technology, RFID has a very promising future. There is little doubt that the technology can bring numerous advantages to these industries. Success in deploying RFID technology, however, will depend heavily on resolving a number of obstacles and impediments before ubiquitous deployment becomes a reality. It is probably fair to say that, at some point, RFID technology will be widely used but it is going to take time. Moreover, while the potential uses of RFID technology may be limitless, it may never reach the expected acceptance level or delivery of its full economic potential due to privacy and ethical concerns, which are discussed later. Despite these caveats, 2005 is the year that the leading global retailers triggered the full-scale propagation of RFID technology.

Adopters of RFID technology can be divided in three categories: early movers, fast followers, and slow adopters.

Early movers are the companies or industries that are leading their industry in terms of RFID adoption and are able to drive major RFID programs that influence their particular industry. They are able to gain the greatest knowledge, have the ability to influence standards, are ready to make significant investments, and take risks.

Fast followers are companies or industries that hesitate to invest in the technology, but aim to gain knowledge and target specific areas at points in time where the cost/benefit can be justified.

Slow adopters are companies or industries that start to implement RFID technology once costs and practices have been stabilized. They will not make any risky investments but are ready to increase speed of implementation based on learning from others in their industry.

3.3.1 A Simplified RFID Technology Roll-out Timeline

In 2004, the number of RFID technology pilot projects by early movers increased rapidly and participants gained experience with the technology. Late in the year, EPC standard Class I Generation 2 was published and European legislation on UHF was amended, solving two important problems.

- **2005**

  EPCglobal becomes fully operational.

  Reliable UHF products become available.

  Vendors offer pallets and crates fitted with RFID tags.

  Early movers, such as Wal-Mart, start large scale roll out throughout the organization, at least at the crate- and -pallet level.

  The number of fast followers starting pilot projects increases quickly.
- **2006**

  EPCglobal standards adopted by ISO.

  Early movers of RFID technology are fully occupied with implementation and system integration.

  Fast followers start their implementation programs.

  Slow adopters of the technology slowly start their initial RFID pilot projects.
- **2007**

  Price of a passive RFID tag continues to fall and begins to approach the 5 cents per tag benchmark price (on large volume purchase).

RFID technology implementation programs of fast followers continues.

Early movers complete their RFID implementation programs with logistical applications.

· **2007 and Beyond**

In the years after 2007, interest will shift towards item-level tagging, but it will be some time before this is implemented. (According to one industry representative, it will be at least 10 years before there is a "no checkout scenario" at large supermarkets. High-value, high-risk goods would be the first to benefit from item-level tagging; goods such as pharmaceuticals and firearms, for example.) Smart shelves for select categories of products begin to appear and "smart" appliances with embedded RFID technology come into the market place.

CHAPTER 4

RFID MIDDLEWARE AND INFORMATION TECHNOLOGY INTEGRATION

4.1 WHAT IS RFID MIDDLEWARE?

In order to reap the full benefits of RFID, those that implement RFID solutions must find ways to incorporate RFID data into their decision-making processes. Enterprise IT systems are central to those processes. Thus, not unless RFID systems are merged into enterprise IT systems will the companies and organizations that invest in RFID be able to improve business and organizational processes and efficiencies.

This is where middleware comes in. Middleware is the software that connects new RFID hardware with legacy enterprise IT systems. PC's ultimately derive their value from the software applications that run on them. In the same vein, RFID hardware is relatively worthless without the software tools that users need to work with it. Middleware is just that: software tools.

Middleware is used to route data between the RFID networks and the IT systems within an organization. It merges new RFID systems with legacy IT systems. It is responsible for the quality and ultimately the usability of the information produced by RFID systems. Some have likened middleware to a traffic cop, in that it manages the flow of data between the many readers and enterprise applications, such as supply chain management and enterprise resource planning applications, within an organization.

RFID-A Guide to Radio Frequency Identification, by V. Daniel Hunt, Albert Puglia, and Mike Puglia
Copyright © 2007 by Technology Research Corporation

4.2   THE RECENT FOCUS ON MIDDLEWARE

Until recently, the whole of the RFID industry's focus lay on tags and readers. As RFID projects have begun to move out of the pilot phase and into the deployment phase, the adopters of the technology are beginning to wonder what they are going to do with all of their new data. Having realized that RFID data is relatively worthless without the software tools needed to manage it effectively, the industry has shifted focus over the last 12 to 18 months to producing middleware solutions.

4.3   CORE FUNCTIONS OF RFID MIDDLEWARE

The term *middleware* has been applied so broadly, not just in RFID but in all of IT, that it has begun to lose real meaning. Unlike other instantiations of middleware, RFID middleware is most often designed to operate at the edge of an IT network rather than close to the center. For example, the middleware components of an RFID network might reside at a factory or at a warehouse, rather than at the center of an organization's IT system. This requires the use of distributed networks and a decentralized IT infrastructure.

RFID middleware moves data to and from points of transaction. For example, in a tag-read process, the middleware will move the data contained on a tag from the reader to the proper enterprise IT system. Conversely, in a tag write process, middleware will move the data from the enterprise IT system to the proper reader and ultimately to the proper tag. RFID middleware has four main functions:

- Data Collection—Middleware is responsible for the extraction, aggregation, smoothing, and filtering of data from multiple RFID readers throughout an RFID network. It serves as a buffer between the volumes of raw data that are collected by RFID readers and the relatively small amount of data that is required by enterprise IT systems in the decision-making process. Without this middleware buffer—parsing through what is important information and what is not—enterprise IT systems could quickly become overwhelmed by the flow of data. For example, it is estimated that when Wal-Mart moves to item-level tagging, it will generate two terabytes of raw data every second.[29]
- Data Routing—Middleware facilitates the integration of RFID networks with enterprise systems. It does this by directing data to appropriate enterprise systems within an organization. In other words, middleware determines what data goes where. For example, some of the data collected by the reader network might be input to a warehouse management

[29] *The Little Chip That Will Change Your Supply Chain Forever*, Microsoft, July 18, 2003.

system to keep track of inventory, whereas other data might be directed to another application to order more stock or debit accounts.

- Process Management—Middleware can be used to trigger events based on business rules. For example, imagine an order is made on a company's website and a pallet is sitting at a dock door in a distant warehouse, waiting for its marching orders. The enterprise IT system responsible for initiating this shipment would pass the purchase order to the middleware system, which would then be able to locate the specific dock door the pallet is sitting at and write the delivery information on its tag. Other events and processes that might be managed by middleware include unauthorized shipment and unexpected inventory, low stock, or stock out.
- Device Management—Middleware is also used to monitor and coordinate readers. A large organization might have hundreds or thousands of different types and brands of readers spread across its network. Networking and monitoring these readers and keeping track of device health and status would be a major job in itself and most efficiently done at the middleware level. Remote management an RFID network could also be made possible through middleware.

4.4 MIDDLEWARE AS PART OF AN RFID SYSTEM—THE EPC ARCHITECTURE

Many of the middleware products currently under development are based on EPCglobal standards, otherwise known as Savant. The Savant specification sorts middleware components according to the functions they serve as shown in Figure 4-1. (EPCglobal standards are discussed later, in greater detail.) There are three functional categories:

- Core Processing
- Reader Interfaces
- Enterprise Application Adapters

4.4.1 Core Processing

Core processing functions sort through and manipulate RFID data collected by a network of readers before directing it to enterprise Information Technology (IT) applications, with the aim of reducing bottlenecks and congestion elsewhere in the enterprise network. Including core processing functions in middleware applications allows data to be handled closer to edge of an enterprise network, such as at a warehouse, rather than at centralized locations, which lessens the burden placed on data transmission networks and central processing computers.

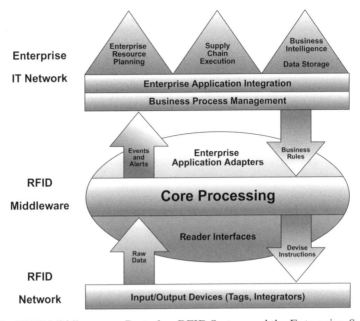

Figure 4-1 RFID Middleware as Part of an RFID System and the Enterprise. Source: Venture Development Corporation.

At a minimum, middleware will eliminate errant, duplicate, or redundant data, which reduces the amount of data that ultimately flows upstream. In a well-tuned operation, most of the data collected by RFID networks will be expected and of little use to enterprise systems. Hewlett Packard (HP), for instance, operates RFID facilities in the United States and Brazil that collectively produce 5 terabytes every day. The company doesn't use all that information, just exceptions and errors in shipments, and middleware applications throw the rest away.

Regarding the downstream flow of information, from enterprise system to RFID network, core processing functions translate business rules and business process management commands into device configuration commands. For example, if an enterprise IT application issues a command to ship a pallet from a West Coast storage facility to an East Coast distribution center, the middleware will be responsible for finding that pallet in its network of readers and keeping track of it throughout the shipment process.

4.4.2 Reader Interfaces

Reader interfaces work between core processing functions and RFID hardware. They are sometimes called edgeware for this reason. They enable RFID systems to discover, manage, and control readers and tags. This can be a difficult task at a large organization with many readers and tags spread throughout large facilities in diverse geographic locations, of potentially differing brands, data formats, and communication interfaces.

Reader interfaces enable uniform communication between core processing middleware functions and readers. Imagine a warehouse that uses Brand X readers at its dock doors and Brand Y readers on its conveyor belts, and that these readers have different data formats and communication interfaces. If this is the case, then somewhere in the communication between the enterprise IT system and the RFID network the distinction has to be made. This is most efficiently done at the middleware level, and more specifically at the reader interface level.

Reader interfaces will enable the communication between core processing and the RFID network by serving as a buffer between the two and shielding one from the other. By doing so, the differences between readers become invisible to every network component upstream, including the core processing functions and the enterprise IT system. The reader interfaces will act as translators between the two sides it stands between, translating the uniform messages issued by the enterprise IT system and core processing components into product specific commands that the Brand X dock door or the Brand Y conveyor belt reader will understand. Conversely, reader interfaces also work in the opposite directing, converting raw data from different readers into a uniform format to be worked on by core processing and ultimately by the enterprise IT system. (An analogy might be the U.N. General Assembly, composed of many ambassadors speaking a number of languages. As the U.N. employs a number of invisible translators to make sure all parties can communicate, RFID reader interfaces would do the same.)

Reader interfaces are also responsible for directing data to the correct reader. For example, in a product recall, an enterprise system might contain the geographic location for a pallet of defective merchandise, but it is unlikely to contain the specific shelf location for that item. The reader interface would be responsible for find the item on the shelf through its network of readers and reporting that information to the proper enterprise application.

4.4.3 Enterprise Application Adapters

Enterprise application adapters work between core processing and enterprise IT systems. They too are a form of edgeware, responsible for delivering RFID data to and from enterprise applications, such as warehouse management, enterprise resource planning, order management, traffic management, and manufacturing execution systems and data warehouses.

Enterprise application adapters convert data flowing out of core processing into application specific events and alerts. Allowing this to occur in middleware at the edge of an enterprise network reduces traffic on enterprise IT networks. In the downstream direction, enterprise application adapters convert business rule and process commands coming from multiple applications in an enterprise system into a uniform format that can be worked on by core processing.

4.5 THE PRESENT STATE OF MIDDLEWARE DEVELOPMENT

Middleware is still in its infancy. Current RFID middleware solutions on the market focus only on reader integration and coordination and basic data filtering abilities. Since most RFID pilot systems have been read-only systems, the middleware solutions available today are read-only also, and have no tag-write functionality built in.

In the future, middleware solutions will have to provide a number of capabilities in order for the organizations that use RFID to reap its full benefits. This includes reader and device management, application integration, partner integration, process management and application development abilities, packaged RFID content, and architecture scalability and administration features.

4.6 MIDDLEWARE VENDORS

Due to the recent interest in middleware, a handful of vendors have emerged, but none dominate yet and their products are still in their infancy. There are several types of players in the market at present:

- Enterprise software application makers offer quick RFID add-ons to existing enterprise software applications and platforms. Supply chain management and warehouse management system providers like Provia, Manhattan Associates, and RedPrairie fall under this category.
- Infrastructure software makers such as Sun, IBM, Oracle, SAP, and Microsoft, are extending existing middleware products to handle RFID. Cisco stated that by 2009, most traffic on Cisco networks will be EPC related and that by 2014, the number of EPC readers worldwide will reach 300 million.[30]
- RFID equipment manufacturers are extending their product lines and entering the middleware market, often through partnerships with other companies. Examples: include Zebra, Check Point, and Intermec.
- Newcomers/startups such as GlobeRanger, OatSystems, ConnecTerra, and Data Brokers offer stand-alone products that filter data and incorporate business rules and task management. Their model for middleware does not require companies to update existing enterprise IT systems, which they believe will allow businesses to leverage prior investments in IT. Their strategy is to push as much of the data processing out to the edge of enterprise networks as possible.

[30] *The Missing Piece*, Peter Winer, Frontline Solutions, July 1, 2004, www.frontlinetoday.com/ frontline/content/printConentPopup.jsp?id=110450.

CHAPTER 5

COMMERCIAL AND GOVERNMENT RFID TECHNOLOGY APPLICATIONS

5.1 INTRODUCTION

The main feature of RFID technology is its ability to identify, locate, track, and monitor people and objects without a clear line of sight between the tag and the reader. Addressing some or all of these functional capabilities ultimately defines the RFID application to be developed in every industry, commerce, and service where data needs to be collected.[31]

The effectiveness of an RFID application in addressing desired functionality is dependent upon several important factors, which include:

- Power—Does the tag contain a built-in power source or can it be only be "passively" activated by the field emitted by the reader? Most applications are currently passive in nature due to cost considerations and passive systems are sufficient for many applications.
- Read Range—Since most RFID is passive, the range of most tags is very limited. This limits the utility of applications to those where assets, merchandise, persons, or animals must be in close proximity to a reader.

---

[31] *RFID—Hot Technology with Wide-Ranging Applications*, David Williams, Directions Magazine, February 25, 2004.

- Storage Capacity—The lowest cost tags have a limited amount of storage capacity (read-only). Recent advancement in RFID technology have increased the capacity and enabled the ability to read/write numerous times, thereby opening up RFID technology to a variety of more dynamic applications.

These factors are basic to any type of RFID application. However, for applications that are very location-tracking or identification oriented, such as in security and access control applications, additional factors need to be considered. These may include privacy concerns and the integration of the RFID technology with other technologies such as Global Positioning Systems and Biometric Technologies.

In the near-term commercial applications of RFID technology that track supply chain pallets and crates will continue to drive development and growth. The RFID industry, as a whole, will focus on applications that rapidly increase volume of use; expecting greater volume will lower costs and the lower costs will accelerate greater demand for development of the technology. Despite the Wal-Mart and DoD mandates, it appears that future market growth for RFID will depend most heavily on the declining cost of the technology. The high cost for the technology, coupled with the additional constraints of the lack of global standards (China) and privacy concerns, act to impede the rapid development of the technology.

While the potential uses of RFID technology have been described as limitless, to date, few "must have" killer application have been fully implemented to spur the explosive development and growth of the technology.

Emerging RFID applications for government use remain in the early adoption phase of development. Recent advances in the scope and breadth of RFID technology development suggests that the technology is beginning to move beyond its traditional commercial boundaries. Manufacturers of RFID equipment are demonstrating that RFID technology works well in many government-related functional environments. At the same time, potential government and public agency users of the technology are being driven to develop a deeper understanding of the benefits and capabilities of RFID deployment in supporting and improving public services.

5.2 EFFECT OF THE WAL-MART AND DEPARTMENT OF DEFENSE MANDATES

RFID technology is being viewed by many in the global economy as a society-changing technology with an unlimited number of potential applications. These visionaries see RFID technology being placed on every object on earth to identify, locate, track, and monitor the object for a variety of purposes.

While RFID technology has been around for a long period of time, its adoption has been uneven, but in June 2004 this dramatically changed when

Wal-Mart announced its mandate to place RFID tags on all shipping containers by January 1, 2005 (see Appendix A, which outlines the Wal-Mart RFID program). Wal-Mart was quickly joined by such other major domestic and foreign retailers as Best Buy, Albertson's, Target, Metro, and Tesco in requiring RFID-enabled deliveries to its distribution centers.

In addition, in August 2004, DoD published its policy guidelines concerning the use of RFID tags on all products within its supply chain and delivered after January 1, 2005 (see Appendix B, which summarizes the key elements of the DoD RFID policy). Wal-Mart and DoD dramatically altered the strategic business landscape for many companies. The expected ripple effect of these dual mandates opened the door for RFID development in smaller companies and industries that, to this point in time, haven't been able to justify deploying RFID technology.

The Wal-Mart and DoD mandates have also generated interest in the development of other RFID applications outside the commercial retail area, such as RFID-enabled personal security and access control devices. Public safety, corrections, and civilian security management-related RFID applications enable comprehensive identification, location, tracking, and monitoring of people and objects in all types of environments and facilities.

The combination of the Wal-Mart and DoD mandates provides formidable champions for the future development of RFID technology. The combined economic strength of Wal-Mart and DoD provides their suppliers, contractors, and vendors little choice but to comply with the mandates. While Wal-Mart and DoD will initially reap the economic benefit from RFID technology deployment, there will also be a tremendous across-industries impact flowing from this deployment of the technology. RFID will ultimately deliver huge efficiencies and cost savings to companies with regional, national, and global supply chains and their customers.

Growth rates for RFID usage range from 35% to 300% per year, with the average in the 40% to 60% range, as shown in Table 5-1. According to IDTech Ex,[32] as of 2004 1.5 billion RFID tags had been sold cumulatively worldwide and this included 500 million active tags. Further, by 2007, the global RFID tag market will be worth about $4.0 billion and will be growing fast. The driving applications for high value use of RFID technology over the next 10 years include those applications shown in Table 5-1.

5.3 STRATEGIC DIMENSIONS OF THE WAL-MART AND DOD MANDATES

There are four strategic dimensions stemming from the Wal-Mart and DoD mandates that will rapidly multiply the number of companies affected by

---

[32] *RFID Explained*, Raghu Das, IDTechEx, 2004.

TABLE 5-1   Potential RFID Tag Sales Volume by Application

Application	Potential Volume	Notes
National ID—contactless smart cards	Italy—50 Million U.K.—58 Million India—500 Million China—970 Million	China issued 8 million contactless cards to their citizens. By 2010, China needs to roll out almost 1 billion cards and many other countries have similar initiatives.
Electronic passports	400 million annually	The U.S., U.K., Thailand, and Australia are among some of the countries progressing to embed smart labels in passports to prevent counterfeiting. First versions are being rolled out now.
Car tires	200 million annually	The TREAD Act in the U.S. is mandating that RFID be used to monitor tire pressure and temperature.
Laundries	Up to 1 billion tags per year (commercial laundries)	Suppliers into laundry applications see this as a very strong growth area. About 70 million tags have been sold to date for this application alone, with many paybacks, including the laundry users being able to use the tag themselves.
Archiving	Up to 100 billion	Potentially a massive market, precursors include library tagging (35 million books to date), event ticketing, and so on.
Conveyances	2 Billion plus per year	Pallets and crates may only demand several billion tags a year, as many are reusable. Therefore, to achieve tens of billions volume demand and the 5 cent tag price as a result, other applications such as high shrinkage items (DVDs, CDs, razors, etc) and airline baggage tagging will also need to grow.

Source: IDTechEx.

RFID technology and further spur demand and the development of new applications[33]:

- Volume and Cost—The volume of RFID tags just from Wal-Mart's top 100 suppliers is estimated at one billion tags per year. With such volume, the cost of RFID tags will begin to fall in 2007. DoD's demand will be even greater. It is estimated that current costs for RFID tags range between 25 and 50 cents and are likely to fall to about 5 cents over the next several years and to 1 cent within 10 years. This cost reduction will

[33] *The Strategic Implications of Wal-Mart's RFID Mandate*, David Williams, *Directions Magazine*, July 2004.

have enormous implications in terms of expansion of RFID technology into new applications area and economic markets.

- Upstream Supply Chain Extensions—As Wal-Mart and DoD suppliers transform to RFID technology, the demand to track products prior to their arrival at distribution centers and military depots will also grow. This includes tracking products once they leave the supplier's shipping area to the time they arrive at their final destination. Requiring suppliers to use RFID technology will also have the likely impact of accelerating the use of RFID technology into the supplier's own supply chain, and eventually, in turn, the supplier's vendor supply chain. This kind of "ripple effect" will greatly multiply the number of companies affected and raise the demand for RFID technology and further lower cost.

- Innovation—As the cost of RFID technology deceases, smaller companies will be able to afford incorporating it into their operations. This will stimulate new kinds of innovative applications and create new markets. Examples of industries that are just beginning to emerge as users of RFID technology include the homeland security industry, which incorporates human and high value asset monitoring and tracking, building/facility access control, identification card management, and counterfeit protection.

- Downstream Supply Chain Extensions—The discussion above mainly addresses business-to-business relationships and transactions, and using RFID technology at the pallet-and-crate level. However, there is also great potential in applying RFID to the individual consumer product for the purposes of managing inventory. As the cost of RFID tags drops closer to the one cent mark, the temptation to apply tags to individual low cost consumer products will grow. The allure of knowing how, when, and from what shelf products are purchased will drive demand for a new generation of merchandising strategies as well as deter shoplifting and employee theft. However, standing in the way of this growth in the use of RFID technology are consumer concerns about personal privacy.

RFID technology is generally viewed by the business community as an evolving technology rather than an immediate ground-breaking technological discovery. There is a reality that the technology is not yet mature enough to accomplish all the processing envisioned to fully exploit all of the potential benefits RFID holds. There are a number of obstacles and privacy impediments that remain to be resolved. As a consequence, RFID technology development is principally being directed to commercial supply chain management applications for which it is now ready, the least disruptive and noncontroversial, and which provides immediate cost savings to companies. RFID technology is making rapid headway in other application areas, such as homeland security and personal security and access control, but these will remain emerging or "niche" markets for the technology's development in the near term.

5.4 RFID TECHNOLOGY FOR BUSINESS APPLICATIONS

While RFID technology has potential applicability in every industry, commerce, or service where data needs to be collected, commercial business sectors will be the focus for growth. These commercial sectors include:

- Transportation and distribution
- Retail and consumer packaging
- Industrial and manufacturing
- Security and access control

As noted, most of the recent growth in RFID technology has been in various commercial sectors of the global economy. However, as potential users of RFID technology gain understanding of the technology and assess the impact it will have on their business environment, user demand will increase and further opportunities for development will emerge. At the present time, new market segments and novel applications remain in the early phase of development and the technology is only just now beginning to move beyond its traditional commercial applications into the government and public agency arena. For the near term, anticipated explosive growth still remains a waiting game: waiting for standards to be implemented, waiting for prices to drop, waiting for major orders to materialize, and waiting for the market to explode with growth. However, with additional strategic thinking and innovative application development, RFID usage will undoubtably expand and become a commonplace technology that is used throughout the world.

Table 5-2 is a list of specific RFID applications prominently in use today. While not necessarily a comprehensive list, it does provide some insight into the scope and breadth of the technology's applicability and usage in today's society. Surely, with the anticipated near-term growth of RFID technology the list is expected to expand rapidly. Further, for ease of readering, the list

TABLE 5-2 Current RFID Commercial Applications

Transportation and Distribution

Fixed Asset Tracking
   Aircraft, Vehicles, Rail Cars
   Containers Equipment
   Real-Time Location Systems

Retail and Consumer Packaging

Supply Chain Management
   Carton Tracking
   Crate/Pallet Tracking

TABLE 5-2   Current RFID Commercial Applications (*Continued*)

Retail and Consumer Packaging

Supply Chain Management
  Item Tracking
  Pharmaceuticals
  Inventory and Tracking

Industrial and Manufacturing

Manufacturing
  Tooling
  Work-in-Progress

Security and Access Control

Pasport and visa management
  Child Tracking
  Animal Tracking
  Airport and Bus Baggage
  Anti-Counterfeiting
  Computer Access
  Employee Identification
  Forgery Prevention
  Branded Replication
  Parking Lot Access
  Room, Laboratory, and Facility Access

Toll Collection

Roads and Bridges

Point of Sale (POS)

Automated Payments
Customer Recognition
Smart Card RFID
Security

Monitoring and Sensing

Pressure, Temperature, Volume, and Weight
Special Facility Access
Facility Security Access
Location within Facility Monitoring

Library Systems

Library Book Collection
Special Collection Access

Source: Technology Research Corporation.

has been linked to the identified commercial sectors of the larger national economy, noted above.

5.5   RFID AND SUPPLY CHAIN MANAGEMENT

The members of a supply chain network—suppliers, manufacturers, and distributors—without any specific effort to coordinate their activities, will operate independently from one another and according to their own agendas. This type of unmanaged network results in inefficiencies, however. The manufacturer might have the goal of maximizing production in order to minimize unit costs, however, if there is not enough demand for product from the distributor, inventories will accumulate. Clearly, all members of a supply chain stand to gain by coordinating their efforts to improve efficiency and overall supply chain performance.

Supply chain management is the combination of process and information technology to integrate the members of the supply chain into a whole. It includes demand forecasting, materials requisition, order processing, order fulfillment, transportation services, receiving, invoicing, and payment processing. Supply chain managers already have many tools to wield at these problem. RFID will be a new tool and will offer an unprecedented ability for supply chain members to coordinate their activities.

In managing supply chains, the information collected by RFID networks will become inputs to the following fundamental decisions[34]:

- Location—of facilities and sourcing points
- Production—what to produce and in which facilities
- Inventory—how much to order, when to order, and safety stocks
- Transportation—mode of transport, shipment size, routing, and scheduling

5.5.1   Supply Chain Metrics

The performance of a supply chain is measured in terms of profit, average product fill rate, response time, and capacity utilization.[35]

- Profit projections can be improved if other parameters are relaxed. For example, by increasing fill time, the costs of transportation could be lowered by making larger shipments. There could be consequences,

---

[34] *Supply Chain Management*, QuickMBA (www.quickmba.com/ops/scm).
[35] *Supply Chain Management and Overview*, (ebusiness.inightin.com/supply_chain/scm_overview.html).

however; customers could be lost if response time is too slow. RFID information, if used to distribute product more efficiently, could lower transportation costs without increasing fill time or lowering profit projections.

- Fill rate can be improved by carrying safety stocks to avoid stock-out. There is obviously a trade-off here between inventory cost and lost profits due to stock-outs. Through improved demand forecasting, enabled by RFID, safety stocks can be lowered without hurting fill rate or risking stock-outs.

- Response time can be lowered at the expense of profits also. There is again a trade-off between the lower costs of long response times and losing customers, particularly in industries where there is a high elasticity of demand; customers might not be willing to wait or might be lost to competition. By using RFID to more accurately target demand and improve order processing and fulfillment, profits can be increased without hurting response time or losing customers.

- Capacity utilization should be high, but not so high that the supply chain can not grow or respond to fluctuations in demand. Problems are often encountered when capacity utilization exceeds 85%. Lower capacity utilization serves as a buffer to enable increased output in the future, should demand rise, whereas high-capacity utilization lowers downside risk through reduced costs. Again, there is a trade-off between the two. By leveraging RFID and improving demand forecasting, materials requisition, order processing, order fulfillment, transportation services, receiving, invoicing, and payment processing, supply chains can lower costs, improve planning for future demand and increase capacity utilization.

5.5.2 Processes of Supply Chain Management

There are five fundamental processes in supply chain management, all of which stand to be improved through the application of RFID technology:

5.5.2.1 Demand Planning and Forecasting   Supply chain software applications use mathematical models to predict future demand from historical data. These models are only as good as the data fed into them. RFID will not only improve the accuracy of data available to the models, but the wealth of the data as well. New mathematical models will be made to make use of the new types of information that RFID systems will produce. More accurate forecasts of demand should result.

5.5.2.2 Procurement   Procurement involves not only the negotiation of prices with suppliers but also the receiving and verifying of shipments. Supply chain management systems enhanced through RFID technology will help to

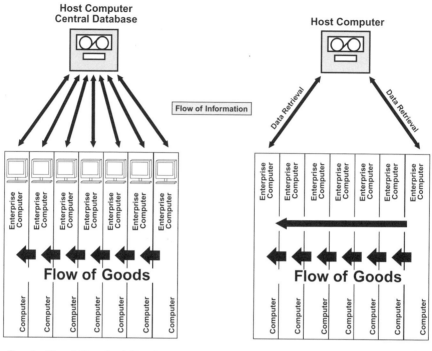

Figure 5-1   Decentralized vs. Centralized Manufacturing. Source: Zebra Technologies ©ZIH Corporation.

further automate these procurement processes, thereby driving the costs of procurement down.

5.5.2.3 Manufacturing and Assembly   RFID can be used on assembly lines to automate and streamline the manufacturing process as shown in Figure 5-1. For example, RFID technology can facilitate a decentralized manufacturing process. In traditional manufacturing processes, all points along an assembly line are networked directly to a central database. As goods flow through the process, the central database has to be updated at each step along the way. This may not always be possible or cost-effective. For example, it might not be possible to network some points in the process to the central database, or it might require too much touch labor to keep track of each item in the pipe.

With RFID technology, a tag can be used as a portable database. Rather than network each point in the manufacturing process to a central database, only a few points in the process need to be connected, such as the beginning and the end. As goods flow through this type of assembly line, important, item-specific information can be recorded directly on the tag. The information would then travel with the item, rather than reside at the central database. Figure 5-1 shows the flow of information in a decentralized versus a centralized computer control system.

To illustrate how the decentralized manufacturing process would work in the real world, consider how a company like Dell Computer might use it. Dell Computer makes PC's to order. When an order is made, the production process begins. Dell allows its customers to choose between a number of options and features when buying their computers, such as processor speed, amount of memory, hard disk size, etc. In a decentralized manufacturing process, all of this configuration information could be written to an RFID tag on a empty case. The empty case would then enter the production process, at which time the central database, which ultimately keeps track of the order, would effectively lose sight of it.

As the case moves through the assembly line, presumably on a conveyor belt, the components would be added as directed by the configuration list and then checked off on the tag as having been installed. At times, the case might come to a junction where computers that need modems are directed left while those that don't are directed right. An RFID reader installed at this point could check the configuration information on the tag and direct the case to the proper point. Eventually, a complete computer, packed inside a box and addressed to the customer that ordered it, would exit the manufacturing process. Assuming this point of the process is networked to the central IT system, the central database could be updated and the distribution process could begin.

5.5.2.4 Distribution Distribution is that portion of the supply chain process where products are delivered to customers and includes warehousing, delivering, invoicing, and payment collection. By automating these processes through the use of RFID, distribution can be made more efficient.

Figures 5-2, 5-3, and 5-4 illustrate how RFID technology could automate the various steps in the distribution process. The first depicts a factory, in this case a pharmaceutical factory. Each item contains an RFID tag. Inside the plant, items can be automatically identified, counted, and tracked. As product leaves the plant, an RFID reader installed in the dock doors checks the contents of the shipment and updates inventories accordingly.

Figure 5-3 depicts a distribution center. When the shipment arrives in the unloading area, RFID readers at the doors examine its contents and update inventories accordingly. The manufacturer can be notified automatically that the shipment has been received and the pallet can quickly be routed to the appropriate delivery truck or to its proper place in the warehouse, all without the need to open packages or examine their contents.

When the shipment arrives at the retail store (Figure 5-4), inventory systems can be updated to include every item. Furthermore, RFID enabled "smart"

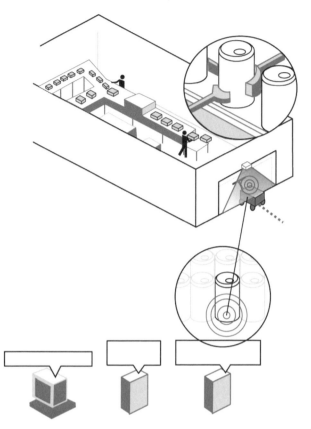

Figure 5-2   Product Leaves the Manufacturer. Source: EPCglobal™.

shelves can automatically order more product from the manufacturer when inventory runs low. This will help to keep stocks at efficient levels. In the future, customers will be able to purchase their items without waiting in check-out line. RFID readers installed at the exits will automatically identify the contents of a customer's basket. Payment could be made without an employee ever having checked the basket, and, in fact, with RFID enabled payment systems, payment could be rendered without the customer even having to break stride on the way to the parking lot. All of this will ultimately lower costs for both customers and retailers.

5.5.2.5 Returns   Returns and refunds are also important parts of supply chain management. RFID can be used to direct defective merchandise back to the manufacturer and more quickly process returns, credit accounts, etc. By automating these processes through RFID the costs of returns can be reduced.

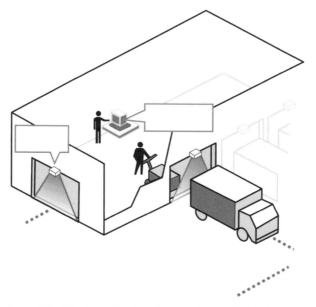

Figure 5-3 The Distribution Center. Source: EPC Global™.

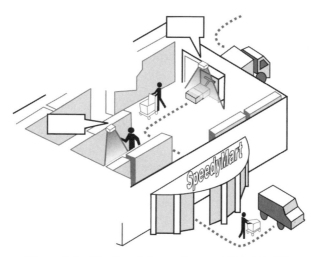

Figure 5-4 The Retail Store. Source: EPCglobal™.

5.6 THE BUSINESS CASE FOR RFID

It is difficult to determine the returns that the early adopters of RFID will see on their investments in RFID tools. Because the technology is still being implemented by the early adopters, and there are still many unknowns, it has not been possible for businesses to accurately quantify the costs involved when evaluating RFID projects for investment, much less the savings or future cash flows that the projects will produce. A survey conducted by ARC of 24 companies actively investing in EPC RFID suggests that they are doing so not because they foresee an attractive return on investment yet but because the Wal-Mart and DoD mandates require them to.[36] In another survey of 80 companies conducted by Accenture, while one-third of respondents expected a high return on investment, fully two-thirds claimed they were still not convinced as to the benefits of RFID.[37]

5.6.1 The Two Types of Return on Investment

The return on RFID investment will come from two sources: direct return on investment (ROI) and ancillary ROI.

Direct ROI will come from optimizing existing business processes. RFID will enable new ways of doing old things. Production and supply lines already in place can be streamlined and made more efficient through the use of RFID technology; labor costs and time to market/warehouse/etc. will be reduced. Quantifying this aspect of ROI is relatively simple and an example will be presented below. These types of returns are more likely to be realized in the short term and require relatively little extra planning to obtain. Some examples, as they pertain to the supply chain, would include better shipping and receiving productivity, improved lot track and trace, improved recall management, and better returns processing.

Ancillary ROI will come from making use of the wealth of information that RFID technology can provide about the systems to which it is applied. RFID will, in this sense, enable businesses to do and know things that have not been possible before. In contrast to direct benefits, the ancillary benefits of RFID will extend beyond the four walls of any individual organization and across multiple organizations to include suppliers and customers.

In quantifying ancillary ROI businesses will need to answer some difficult questions, like what it's worth to know in real time where every truck in a supply line is or what's the value of knowing where every pallet is in a warehouse and how long it's been there? The answers to these questions ultimately lie in how well businesses utilize RFID data.

The ancillary benefits of RFID will be realized in the long term and require more planning and critical thought and business analysis to obtain. Some examples of ancillary benefits, as they pertain to supply chain planning, would include: reduction in inventory and working capital, improved revenue through reduction in stockouts, and reduced expediting costs.[38]

---

[36] *Return on Investment Is Lacking for EPC RFID*, Steve Banker, ARC Advisory Group, 2004.
[37] *High Performance Enabled through Radio Frequency Identification—Accenture Research on Manufacturer Perspectives*, Accenture, 2004.
[38] *High Performance Enabled through Radio Frequency Identification—Accenture Research on Manufacturer Perspectives*, Accenture, 2004.

5.6.2   A Short-Term Focus

Due to the uncertain long-term ROI and the immediacy of the Wal-Mart and DoD mandates, for now manufacturers seem to be focusing on the short-term benefits of RFID, which mostly fall under the direct ROI category. In Accenture's survey, companies were asked to rate the many potential benefits of RFID on a scale of 1 to 5 in increasing benefit. The top-rated categories were those that would fall under the direct ROI column. The top three were:

- Improved lot track and trace
- Improved recall management
- Better shipping and receiving

From this it would appear that many long-term, ancillary benefits are not intuitively recognized and have not been considered by the businesses investing in RFID.

5.6.3   Quantifying Return on Investment

5.6.3.1   Example—An ROI Study Conducted by RFID Wizards Inc.[39]

RFID Wizards acquired by Traxus Technologies Inc., has published an ROI study for RFID solutions. In the study, a hypothetical manufacturer and a hypothetical store distribution center form a simple supply chain. The companies use different IT systems and different sets of product numbers however. When pallets are delivered from the factory to the distribution center, the product numbers are translated manually and entered into the IT system. The report proposes an RFID solution for the distribution center warehouse, enumerates the effects it will have on warehouse processes, and calculates the payback period, or the number of months until the initial RFID investment will have been recovered through savings.

Cost of Investment   The report breaks down the cost of installing an RFID system at the warehouse is shown in Table 5-3.

RFID tags are another cost of investment. The total cost of RFID tags will depend upon the number of pallets shipped per month. In the report, it is assumed that RFID tags cost $0.85 per pallet.

Savings and Reduced Costs of Labor   The report also breaks down how the RFID system will affect various warehousing processes, minute by minute and worker by worker. It concludes that in total, 30 worker-minutes per pallet will be saved by automating the data entry process through the use of RFID. Assuming an average employee cost for a warehouse worker and inventory

---

[39] *Return on Investment Study for RFID Solution*, RFID Wizards Inc., 2003.

TABLE 5-3   Breakdown of Cost for Installing RFID System in Warehouse

Equipment	Cost	Qty	Total
RFID Enabled Dock Door	$8,000.00	4	$34,000.00
RFID Host Server	$3,000.00	2	$6,000.00
Handheld RFID/Barcode Reader	$2,199.00	4	$8,796.00
Misc. Hardware & Cables	$500.00	1	$500.00
2-Year Warranty	$9,759.20	1	$9,759.00
			$59,055.20
	Consulting Services		
RFID Engineer (person days)	$1,200.00	75	$90,000.00
			$90,000.00
Total Equipment and Services			**$149,055.20**

Source: RFID Wizards Inc.

control clerk of $18.20/hour ($14.00/hour + 30% overhead), this produces a savings of $9.10 per pallet.

Payback Period   Finally, the report calculates the accounting payback period. Straight-line equipment depreciation over 36 months with $0 recovery value is used. If the warehouse processes an average of 500 pallets per month, the payback period turns out to be 26 months. The payback period is shorter if more than 500 pallets per month are transferred between the facilities. The conclusions of the report are summarized in Figure 5-5.

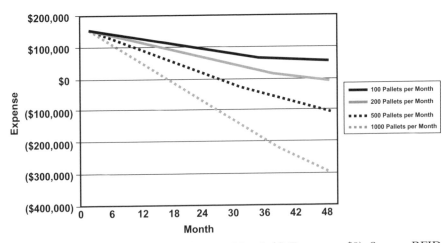

Figure 5-5   Payback Period vs. Pallets per Month (@ Expense = $0). Source: RFID Wizards Inc.

5.6.3.2 *Example—ARC Advisory Group Emerging Practices Study*[40]

The 2004 survey by ARC could cast suspicion over the 1,000 pallets per month payback period in the graph shown in Figure 5-5; it seems a bit more optimistic than the view held by many of the companies who have already invested in RFID. Of all respondents to ARC's survey, 95% believed that the payback period would be greater than two years. Still, RFID Wizard's exercise serves as a good introduction to quantifying RFID return on investment.

In ARC's emerging practices study, the 24 respondents, on average, claimed first-year costs of their projects to be $11.5 million and first-year savings to be $1.5 million, which would put the payback period at much greater than two years.[41]

5.6.4 Return on Assets (ROA)[42]

When companies are unable to find profitable projects to invest in, or when economic downturns strip them of the capital to do so, they will focus their efforts on obtaining a higher return on existing assets by improving asset management processes. RFID lends itself to this problem.

Effective asset management programs ensure that workers always have access to the tools, equipment, and resources they need, when and where they need them. There are two ways of accomplishing this: by either tightly controlling assets through record keeping and control procedures, or by maintaining spare resources as a safety stock. RFID technology makes the record-keeping option much less expensive and labor intensive than it has been until now.

The time spent by employees on equipment searches can be costly. In a survey by WhereNet, a wireless asset management systems vendor, 64% of companies who responded stated that their operations personnel conduct equipment and inventory searches every day, while 27% claimed they make up to 10 such searches a day. Furthermore, nearly half of the respondents claimed the searches take up to one hour. Even if an employee spends 10 minutes per day on equipment and inventory searches, over the course of a year this adds up to a 40 hour work week.

Equipment loss through misplacement, theft, or employee "borrowing" can be even costlier. Take for instance a $60 cordless drill that is taken home by an employee for short-term personal use. If another operations employee,

---

[40] *Return on Investment Is Lacking for EPC RFID*, Steve Banker, ARC Advisory Group, 2004.

[41] *RFID: ROI More than 2 Years*, from the pages of Control Engineering (www.manufactureing. net/ctl/index.asp?layout=articlePrint&articleID=CA477152), November 11, 2004.

[42] *Increasing Profits and Productivity: Accurate Asset Tracking and Management with Bar Coding and RFID*, Zebra Technologies.

paid at $18/hour, spends 30 minutes looking for the drill, it will have cost the company $9. If the supervisor, paid at $30/hour, spends another 10 minutes on a cursory search, it will have cost the company $14. If the drill is replaced for $60, the total will have reached $74.

For a company that earns the S&P average after-tax profit of 10.72%, $690.30 of revenue are needed to replace the drill. If the operations employee is sent to a hardware store to buy the drill, the required revenue climbs to $858. Finally, the original task the employee needed the drill for won't be completed as scheduled.

From this is evident that costs of poor asset management can add up quickly. The long-term business case for RFID technology, though still difficult to quantify accurately, is not difficult to make. By RFID tagging even relatively inexpensive assets like cordless drills and installing a location identification RFID system, companies can eliminate costly equipment searches and the losses due to theft and borrowing.

5.6.5   The Routes to Return on Investment

Ultimately, to make a business case for RFID, it has to be shown that RFID will increase shareholder value. This can come about through three avenues: increase in revenues, increases in operating profits, and increases in capital efficiency. Figure 5-6 shows several routes to obtain a return on your investment in RFID.

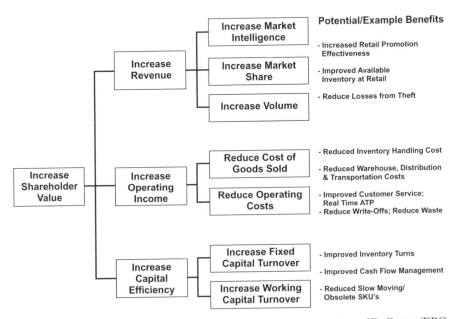

Figure 5-6   A Breakdown of Return on Investment. Source: Auto-ID Center/EPC global.

5.7   GOVERNMENT USE OF RFID TECHNOLOGY

Federal, state, and local governments are taking a larger role in the deployment of RFID technology. DoD is currently the leader in government use of RFID technology and is engaged in developing innovative uses of the technology from tracking items within its supply chain to tracking armaments, food, personnel, and clothing to war theaters.

Other federal agencies are rapidly following suit with their own RFID projects:

- The General Services Administration (GSA) has mandated the use of RFID to help it manage information on the buildings, fleets of cars, and supplies it manages.
- The Department of Homeland Security's (DHS) NEXUS program uses RFID in low-risk traveler card for U.S. Canada border crossings (approximately 50,000 enrollees).
- The U.S.-VISIT "next generation travel documents" program is designed to record the entry and exit of non-U.S. citizens to the United States, verify the identity of incoming visitors, and confirm compliance with visa and immigration policies. The system is designed to record the entry and exit of visitors through the use of digital finger scans and digital photographs captured at the port of entry.
- The Department of State recently announced that the cover of the U.S. passport will be embedded with an RFID chip. The RFID passport was issued to officials and diplomats in early 2005 and to the public by the end of 2006. Each passport will contain an electronic version of the same personal data as now appears on the inside pages of the passport. In addition, a digitized version of the photograph, holograms, security ink and a "ghost photo" will be used. The passport will be read remotely with an RFID reader.
- DHS is also pushing for the adoption of RFID for cargo containers to determine if a container has been tampered with prior to its entry into the country.
- The Food and Drug Administration (FDA) is looking at requiring manufacturers to embed RFID tags into pharmaceutical labels. The aim is to be able to find exactly where the drug is on the shelf and how long it has been there. This information would also be useful for drug recalls or to more effectively verify expiration dates and prevent counterfeiting.
- The U.S. Postal Service is considering putting RFID tags on postage stamps to better track and locate mail more quickly.
- The Internal Revenue Service and the European Union are interested in exploring the use of RFID in money to prevent counterfeiting.
- Airports, which are generally run by state, county, or municipal governments, are interested in deploying RFID technology to ensure total,

all-around security at airports within their jurisdictions. The airline industry has begun using RFID tags to screen, sort, and reduce labor and
maintenance costs in their baggage-handling operations.

• In Buffalo, New York, a private elementary school has begun to record
the time of day students arrive in the morning using RFID. Eventually,
the school plans to use the RFID system to track library loans, disciplinary records, cafeteria purchases, and visits to the nurse's office.

Although RFID technology is still in its infancy, Table 5-4 provides a list of
recent examples of RFID applications that are currently being used by federal,
state, and municipal governments.

Clearly, government use of RFID technology is picking up momentum and
government agencies, at all levels, appear to want to fully exploit RFID technology. This has some privacy advocates concerned. To them, there has been
no serious public policy debate on the privacy implications and limitation of
the proper uses of RFID technology by government agencies. They claim their
concerns are particularly acute in that private citizens have no other option
to receiving public services elsewhere. Consequently, they may have to com-
promise personal privacy to receive government services.

In the view of the privacy advocates, the development and growth of RFID
technology is not a purely commercial market situation, determined by the
laws of economic supply and demand. The use of RFID by the government,
in effect, creates an enormous government subsidy for RFID technology development by generating economies of scale for the production of tags, readers,
and other RFID equipment. This will drive down the cost of the technology
and further expand and legitimize its use. They claim that widespread

TABLE 5-4   Federal, State, and Municipal Applications of RFID

Location	Government Entity	Use	Notes
AZ	Prisons	Inmate Tracking	
AZ	U. of Arizona	Parking Permits	
AZ	Calipatria State Prison	Inmate and Guard Tracking	
CA	Caltrans	Bridge Toll Payments	
CA	Public Libraries, S.F., Berkeley, Santa Clara	Tagging Library Collection to Facilitate Management	
CA	UCLA	"Smart Kindergarten Project" Assessment of Student Collaboration in Small Group Settings	Not RFIDs, per se, but sensors that measure location, orientation and speech

TABLE 5-4   Federal, State, and Municipal Applications of RFID (*Continued*)

Location	Government Entity	Use	Notes
CA	Building Commission	Inspecting Elevators and Amusement Rides	
ME	The Lobster Conservancy	Lobster Tracking	
NV	McCarron International Airport, Las Vegas	Tracking Passenger Bags	
TN	Oak Ridge National Laboratory	Evacuation Monitoring and Evacuation Systems (EMAS)	
TX	Harris County Toll Road Authority; North Texas Tollway Authority	Toll Collection	
US	CBP	Free and Secure Trade Program (FAST)— Border Crossing System and "E-Seals"	
US	DHS	US-Canada Border "NEXUS" Program	
US	DOD	Use of Tags on Pallets from Manufacturers	Mandated 10/03
US	DOD, etc.	Coordinating Various Agencies	
US	FDA	Pharmaceutical Labels to Find a Drug on a Shelf and How Long it Has Been There	Under Consideration
US	Federal Highway Administration	Highway Safety	
US	GSA	To Manage Information on Vehicles, Building, etc.	Use Mandated
US	IRS	Putting Chips in Money to Combat Counterfeiting	Under Consideration
US	INS	Automated Border Crossing	
US	Military	JTAV	
US	USDA	Livestock Tagging	Plans to Develop, but Lack of Funding
US	USPS	Chips on Stamps to Track Mall	Under Consideration
WA	Seattle Public Library	Tagging Collection to Facilitate Management	

government use of RFID will result in the development of even more RFID sensors, individual databases, and personal record keeping capabilities, which will eventually be woven into the fabric of the social environment and threaten individual privacy.

5.8   RFID AND THE PHARMACEUTICAL SUPPLY CHAIN

It is estimated that 7% of all medications in the international pharmaceutical supply chain are counterfeit.[43] In some countries the problem is endemic and patients are more likely to receive fake drugs than real. Though counterfeiting in the United States has been kept comparatively low through the establishment of a comprehensive system of laws, regulations, and enforcement at both the federal and state levels, the FDA has seen a rise in drug counterfeiting cases in recent years. The number of counterfeit drug investigations conducted by the FDA averaged 5 throughout the 1990s, however, that number has risen to over 20 per year since 2000,[44] as shown in Figure 5-7. The full extent of the problem is much worse than the trend indicates; in one such case, in 2003, over 200,000 bottles of counterfeit Lipitor found their way into U.S. markets.[45]

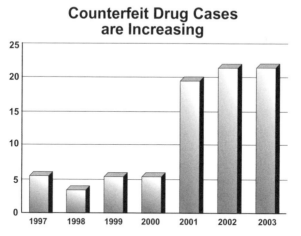

Figure 5-7   FDA Counterfeit Drug Cases per Year. Source: The Food and Drug Administration.

[43] *Item-Level Visibility in the Pharmaceutical Supply Chain: A Comparison of HF and UHF RFID Technologies*, Philips, *et al*, July 2004.
[44] *Combating Counterfeit Drugs*, The Food and Drug Administration, February 18, 2004.
[45] *Tiny Antennas to Keep Tabs on U.S. Drugs*, Harris, Gardiner, *The New York Times*, November 15, 2004.

In response to the growing problem, the FDA formed the Counterfeit Drug Task Force in 2003. That group received comments from a wide range of security experts, federal and state law enforcement officials, technology developers, manufacturers, wholesalers, retailers, consumer groups, and the general public on ways to prevent or deter drug counterfeiting. Using these ideas, the FDA developed a framework for the pharmaceutical supply chain to combat modern counterfeiting threats. That framework was described in a February 18, 2004, report entitled "Combating Counterfeit Drugs." Though their approach to addressing the problem of counterfeiting has been multipronged and includes a number of technology, policy, and legal options, RFID technology has been identified as "the most promising approach to reliable product tracking and tracing."

5.8.1   A Summary of "Combating Counterfeit Drugs"

In its report, the FDA set several goals and a timeline for the pharmaceutical industry to implement track-and-trace anti-counterfeiting measures. As regards to RFID, the report claims:

> The adoption and common use of reliable track and trace technology is feasible by 2007, and would help secure the integrity of the drug supply chain by providing an accurate drug "pedigree," which is a secure record documenting the drug was manufactured and distributed under safe and secure conditions.

Their ultimate goal is to make the copying of medications either extremely difficult or unprofitable for would-be counterfeiters through the use of RFID technology. It is generally viewed that the use of mass serialization to uniquely identify the contents of the U.S. drug supply chain at the pallet, case, and package levels (such as with EPC labels) will accomplish this goal.

Studies are underway to evaluate the feasibility of the late 2007 timeline. Due to the high price of pharmaceuticals, it is believed that the cost of implementing RFID technology on so short a schedule will be offset by the cost-saving benefits that RFID will also provide to the industry. Below is the timeline for the industry's adoption of RFID technology as seen by the FDA and outlined in the report:

- **2004**

    Performance of mass serialization feasibility studies using RFID on pallets, cases, and packages of pharmaceuticals
- **2005**

    Mass serialization of some pallets and cases of pharmaceuticals likely to be counterfeited

    Mass serialization of some packages of pharmaceuticals likely to be counterfeited

Acquisition and use of RFID technology by some manufacturers, large wholesalers, some large chain drugs stores, and some hospitals

- **2006**

Mass serialization of most pallets and cases of pharmaceuticals likely to be counterfeited and some pallets and cases of other pharmaceuticals

Mass serialization of most packages of pharmaceuticals likely to be counterfeited

Acquisition and use of RFID technology by most manufacturers, most large wholesalers, most chain drugs stores, most hospitals, and some small retailers

- **2008**

Mass serialization of all pallets and cases of pharmaceuticals

Mass serialization of most packages of pharmaceuticals

Acquisition and use of RFID technology by all manufacturers, all large wholesalers, all large chain drugs stores, all hospitals, and some small retailers

The report also identified the goal of meeting and surpassing the requirements of the Prescription Drug Marketing Act (PDMA) of 1987 through the use of RFID. The PDMA requires the pharmaceutical industry to establish a pedigree system to track and trace pharmaceuticals in the U.S. drug supply. At the time the legislation was passed, the technology to do so electronically did not exist and so the industry has had to contend with implementing a paper pedigree. This has posed practical and administrative difficulties. Implementation of the pedigree regulation has been delayed.

Some in the pharmaceutical industry have also expressed doubts about a paper pedigree's usefulness or feasibility. Aside from the high cost of implementing a paper pedigree, it is feared that paper pedigrees are too easy to forge and counterfeit.

RFID technology is seen as a solution to these problems. By the time the paper pedigree system has been fully established, the means to implement a fully electronic pedigree system through RFID will exist (and perhaps even at a lower cost). Furthermore, an electronic RFID pedigree system will not only meet but surpass the requirements of the PDMA. As a result, the pharmaceutical industry and the FDA are seeking to meet the requirements of the PDMA through the use of RFID.

5.8.2 The Follow-up to "Combating Counterfeit Drugs"

In November of 2004, the FDA published a Compliance Policy Guide (CPG) for implementing RFID feasibility studies and pilot programs. This report was the FDA's first step in implementing the recommendations outlined in "Combating Counterfeit Drugs." At the time of publication, Wal-Mart, Accenture, and CVS drugstores had already begun feasibility studies and pilot programs. Furthermore, concurrent with publication of the initiative, the following was announced:

- Purdue Pharma announced that it would begin tagging shipments of OxyContin headed to Wal-Mart stores and wholesaler H.D. Smith at the case level. OxyContin is a Schedule II controlled substance that has become the target of much theft in recent years and is subject to wide-spread abuse.
- Pfizer announced that it planned to begin tagging all bottles of Viagra intended for sale in the United States by the end of 2007. Viagra is one of the most counterfeited drugs in the world and the United States.
- GlaxoSmithKline announced it intends to begin using RFID on at least one product deemed as susceptible to counterfeiting by the National Association of Boards of Pharmacy. Possible candidates include several HIV medications.
- Johnson and Johnson claimed that it too was involved in pilot studies, without naming any specific medications.

The FDA has issued the CPG with the belief that it will clear the way for more pilot programs to begin, especially for those drugs that are likely targets of counterfeiting.

5.8.3   The Pharmaceutical Industry Embraces RFID

While the FDA is interested in RFID primarily as a means of securing the nation's drug supply, the pharmaceutical industry has embraced the technology for other reasons as well.

If 7% of all drugs in the international supply chain are counterfeit, there can be no doubt that counterfeiting has had a marked impact on the industry's profitability. Furthermore, it has been estimated that 6–10% of the U.S. drug supply is stolen or diverted at the retail level. Then, "gray market" distribution, in which pharmaceuticals are diverted to unauthorized channels, is estimated to cost drug companies hundred of millions of dollars a year also.[46] RFID technology and item-level visibility of the supply chain could significantly reduce all of these problems and the costs associated.

The pharmaceutical industry could benefit from using RFID technology in other ways, too. Every year more than $2 billion worth of overstocked drugs are returned, having expired on the shelf and gone to complete waste. As with other industries, item-level visibility of the supply chain can help to eliminate this inefficiency.

---

[46] *Item-Level Visibility in the Pharmaceutical Supply Chain: A Comparison of HF hand UHF RFID Technologies*, Philips, *et al*, July 2004.

In 2001, the pharmaceutical industry issued 1,300 product recalls.[47] The time spent identifying and locating the products recalled poses another high cost to drug companies. Mass serialization will make it easier to zero in on specific recall targets, thus reducing the amount of time and money spent on recall efforts, as well as the likelihood of mass market recalls.

The sum of all savings to the pharmaceutical industry as a result of implementing RFID-based track-and-trace solutions amounts to more than $9 billion by 2007 alone, it is estimated, with much greater savings to come in the years following.

5.9 RFID IMPLANTED IN HUMANS[48]

In October 2004, the FDA approved Applied Digital Solution's (ADS) plans to market a subdermal implantable microchip that provides RFID access to an individual's medical records. Approval came after review of the privacy and confidentiality issues that could arise from the implanted device. The device is designed to allow doctors to scan patients to ensure positive identification and that they receive the proper treatment and medications.

Applied Digital Solutions claims the device, called VeriChip, will save lives and reduce injuries due to errors in medical treatment. Privacy advocates fear that the device's approval will lead to the eventual tracking of people through implanted RFID devices. Some fundamentalist religious groups have objected to the device's approval also.

According to Applied Digital Solution's marketing materials, the VeriChip has a variety of uses. Company literature describes it as "a miniaturized implantable radio frequency identification device that has the potential to be used in a variety of security, financial, and other applications." The implants are about the size of a grain of rice, with a unique verification number, which is captured through the use of a proprietary reader. The company is also attempting to develop an implant that would contain GPS capabilities as well, which would allow the implanted chip to be pinpointed anywhere on the globe. The device had already been approved for use in livestock and pet identification applications before being approved in humans. Millions of animals in recent years have had the device embedded.

In other countries, prior to its approval in the United States, VeriChip was already being marketed for use in humans.

The device is being used in Mexico as an anti-kidnaping measure. Mexico suffers from a kidnaping epidemic, with up to 3,000 people abducted every year. Thousands of Mexican citizens recently demonstrated to demand gov-

---

[47] *Item-Level Visibility in the Pharmaceutical Supply Chain: A Comparison of HF and UHF RFID Technologies*, Philips, *et al*, July 2004.
[48] *U.S. Agency Clears Implantable Microchips*, Barnaby J. Feder and Tom Zeller, Jr., *The New York Times*, October 15, 2004.

ernment action to end the kidnapings. (It should be noted that, in the past, the biggest security problem for Mexican law enforcement has been corruption by law enforcement officials themselves and their suspected involvement in many of the kidnappings.) The Mexican distributor of VeriChip claims that about 1,000 Mexicans have been implanted with the device for this purpose.

Furthermore, in the summer of 2004, the attorney general of Mexico announced that he and many of his subordinates had been implanted with the chip. It was not a medical or anti-kidnaping application of the technology, however, but rather a security application. The device in this case is being used to control access to secure rooms used to store documentation relating to the prosecution of Mexico's drug cartels.

In Europe, the owner of two clubs in Spain and the Netherlands offers implantable chips to patrons whom wish to dispense with traditional membership, identification, and credit cards. Club patrons can have the chip implanted in their arm or hand. Once the chip is implanted, the club patron can pay for drinks with a wave of the hand. Access to special VIP sections of the club can also be obtained without resorting to badges or other means of identification. So far, about 150 people have received the chip.

RFID implant manufacturer VeriChip has announced that 280 patients from the New Jersey area are to be have health records chips inserted under their skin as part of a trial into the use of the technology to manage long-term conditions.

Volunteers who are patients of the Hackensack University Medical Center, Hackensack, NJ and suffer from chronic heart disease, epilepsy, diabetes or are recent recipients of organs, will have the RFID chips, the size of a grain of rice, implanted above their right elbow.

The passive chips will contain a 16-digit number that, when scanned at the medical center, will link them to their electronic patient record. Patients who present at the emergency room who are unable to identify themselves or provide their medical history are expected to benefit.

The chips themselves will not contain the records, but the 16-digit number obtained by reading the chip with an RFID reader can be linked to the existing health records at the centre. These will contain family contact information, recent lab test results, pharmacy prescription information and medical information from the records of Horizon Blue Cross Blue Shield of New Jersey (HBCBSNJ), the health insurer that is carrying out the trial.

The chips will be provided free-of-charge to patients who sign up to the project, which is being funded by HBCBSNJ. It was not clear at time of publication whether the trial results will be published, but the insurer will use the results of the trial to see whether it should be expanded.

CHAPTER 6

RFID TECHNOLOGY IN HOMELAND SECURITY, LAW ENFORCEMENT, AND CORRECTIONS

6.1 INTRODUCTION

Many large companies and industries, particularly retail, manufacturing, transportation, and logistics, are rapidly adopting or being driven to RFID technology. These companies and industries have come to recognize that there are real economic efficiencies and payoffs that can be gained through the deployment of RFID, particularly in supply chain and asset management operations.

Business interest in RFID technology is driven by a desire to achieve greater speed and visibility into supply chains, with the goal of increasing both operational efficiency and individual store effectiveness. An efficient supply chain ensures that goods are delivered to the right place and at the right time when consumers are ready to purchase. This also ensures lower inventory levels, reduced labor costs, and increased sales for the business.

No other technology has proliferated into the business mainstream as rapidly as RFID, and the rapid technological advances surrounding RFID have significantly enhanced its flexibility and adaptability so that it can now be easily applied to any public sector enterprise or operation as well. Moreover, many of the economic benefits and attributes surrounding RFID technology also apply and are desirable in public sector enterprises and agencies.

RFID-A Guide to Radio Frequency Identification, by V. Daniel Hunt, Albert Puglia, and Mike Puglia
Copyright © 2007 by Technology Research Corporation

While private sector supply management operations are driving current RFID development, innovative uses of RFID technology are beginning to emerge and promise much of the same benefit and value in the public sector as well.

RFID technology holds exciting opportunities for almost any enterprise and can deliver real value when applied to a well-defined and controlled process, and this would include homeland security, law enforcement, and corrections operations. Accordingly, the following sections will review some of the recent homeland security, law enforcement, and corrections deployments of RFID technology, recognizing that the technology is only now beginning to be integrated into the government's infrastructure but will probably expand rapidly.

6.2   RFID TECHNOLOGY IN HOMELAND SECURITY

The nation's advantage in science and technology is a key ingredient to improving homeland security. The national vision for science and technology in homeland security calls for the Department of Homeland Security (DHS) to be a focal point for a national research and development enterprise, similar in emphasis and focus to that which has supported the national security community for more than fifty years. The national vision for homeland security science and technology development states:

> The Department of Homeland Security (DHS) will establish a disciplined system to guide its homeland security research and development effort and those of other departments and agencies. As a Nation, we will emphasize science and technology applications that address catastrophic threats. We will build on existing science and technology whenever possible. We will embrace science and technology initiatives that can support the whole range of homeland security actors. We will explore both evolutionary improvements to current capabilities and development of revolutionary new capabilities. DHS will ensure appropriate testing and piloting of new technologies. Finally, DHS, working with other agencies, will set standards to assist the acquisition decisions of state and local governments and private-sector entities.

RFID technology development and its identification and location determination capabilities falls within the scope of the national vision for homeland security science and technology. One of the major national homeland security science and technology initiatives indicates:

> Apply biometric technology to identification devices—These challenges require new technology and systems to identify and find individual terrorists. The Department of Homeland Security would support research and development efforts in biometric technology, which shows great promise. The Department would focus on improving accuracy, consistency, and efficiency in biometric systems.

As a technological solution to a complex and far reaching problem, RFID technology is well suited to improving homeland security. It has many inherent qualities and capabilities that support (1) identity management systems and (2) location determination systems that are fundamental to controlling the U.S. border and protecting transportation systems.

RFID technology can also be combined with other technologies, i.e., smart card, GPS, and communications and information systems, and the data gathered through deployment of all these technologies can be used to support various homeland security intelligence gathering functions.

Clearly, the fusion of biometric identification and location determination systems with RFID technology is a driving force to improving homeland security. One only has to look at the border and transportation security goals of the National Strategy for Homeland Security to foresee that this fusion will be a key component to securing U.S. borders and national transportation systems. Two of the major initiatives of the border and transportation security strategy that will require extensive use of RFID technology are:

- Create "Smart Borders"—At our borders, the DHS could verify and process the entry of people in order to prevent the entrance of contraband, unauthorized aliens, and potential terrorists. The DHS would increase the information available on inbound goods and passengers so that border management agencies can apply risk-based management tools. It could develop and deploy required entry-exit system to record the arrival and departure of foreign visitors and guests. It could develop and deploy non-intrusive inspection technologies to ensure rapid and more thorough screening of goods and conveyances. And it could monitor all our borders in order to detect illegal intrusions and intercept and apprehend smuggled goods and people attempting to enter illegally.

- Increase the Security of International Shipping Containers—Containers are an indispensable but vulnerable link in the chain of global trade; approximately 90% of the world's cargo moves by container. Each year, nearly 50% of the value of all U.S. imports arrives via 16 million containers. The core elements of this initiative are to establish security criteria to identify high-risk containers' pre-screen containers before they arrive at U.S. ports; use technology to inspect high-risk containers; and develop and use smart and secure containers.

The terrorist attacks of September 11, 2001 (9/11) provided a strong impetus for RFID technology development in the United States. The 9/11 attacks produced a trend away from a "reactive" to a more "proactive" approach to homeland security protection. The DHS has clearly embraced RFID technology as one of the technologies of choice to improving security protection at the borders and ports of entry to the United States.

Example of new RFID technology applications under current development that support U.S. border and port of entry security protection include:

- Vehicle, driver, passenger, and personal identification border crossing systems
- Vehicle registration systems
- Access control for vehicles in a gated environment
- Revenue control, payment, and tracking systems with an audit trail
- Imported goods traceability and security systems
- Container tracking and tracing systems
- Air cargo, baggage, and passenger control programs

RFID technology is widely expected to improve homeland security through progressive program development and on-going applications deployment. DHS has initiated the RFID technology program through the U.S.-VISIT initiative, which operates at 115 airports and 14 seaports. U.S.-VISIT combines RFID and biometric technologies to verify the identify of foreign visitors with non-immigrant visas. With this technology, digital fingerprint and digital photographs are recorded and terrorist watch lists are checked to make sure potential terrorists don't enter the country.

Successive steps to deploying RFID technology in homeland security include:

- Installing biometric equipment and software at all ports of entry so biometric passports can be used.
- Requiring foreign visitors with visas to do a finger scan as they leave the United States, not just when they arrive in the country. This initiative has been tested at Baltimore-Washington International Airport using kiosk scanners.
- Expanding the Use of Smart Cards—A credit card size plastic card with an embedded computer chip that can be either a microprocessor or a memory chip. The chip connection is either via direct physical contact or remotely via an electromagnetic interface.
- Creating a Transportation Worker Identification Credential—A smart card issued to public or private employees who have access to secure areas of ports, railways, and airports.
- Authenticating the identity of airline passengers by checking their records (name, address, and date of birth) against commercial databases and terror watch lists. It should be noted that the Government Accounting Office is studying the privacy implications and effectiveness of this initiative before Congress gives its approval.
- Tracking cargo and cargo security is a huge concern for DHS. The U.S. Customs Service initiated the Container Security Initiative (CSI) in FY 2002 to extend security beyond the immediate U.S. border by identifying and examining cargo containers before they are shipped to the United States. In conjunction with the CSI program, the "Smart & Secure Trad-

elanes" (SST) program is a phased, industry-driven initiative that provides RFID smartseals, readers, and other sign posts for container identification and tracking using RFID technology. This homeland security RFID application will provide instant notification of container security breaches.

RFID technology makes immediate economic sense in areas where the cost of failure is great. Homeland security is one area where a high premium can be placed on preventing problems before they occur. RFID technology, as an enabling technology, is an ideal means of locating, tracking, and authenticating the movements of people and objects as they enter and depart the United States.

For the foreseeable future, developing effective homeland security RFID applications will continue to be a major stimulus and driver in RFID technology development.

6.3   RFID IN LAW ENFORCEMENT

Adoption of RFID technology for law enforcement applications has been slowly developing in the United States, principally due to privacy concerns and lack of awareness in the potential of the technology to improve law enforcement operations. Law enforcement in the United Kingdom, on the other hand, has been much more aggressive in developing and deploying RFID applications in police operations, particularly in the areas of traffic management and property crime prevention. There does not appear to be the same level of public concern about personal privacy considerations surrounding RFID technology in the United Kingdom as in the United States.

RFID technology applications for law enforcement operate in three dimensions:

- Applying RFID Technology to Improve Police Efficiency
- Applying RFID Technology to Ensure Police Officer Safety
- Applying RFID Technology as a Crime Fighting Tool

6.3.1   Applying RFID Technology to Improve Police Efficiency

Law enforcement and other public organizations are always under pressure to better manage their operations, reduce costs, and improve service. Based on the widespread private sector enthusiasm for RFID technology development, there is also the expectation that an operational improvement and efficiency case can be made for deploying RFID in law enforcement agencies. By automating and improving law enforcement business and administrative processes through the application of technology, such as RFID, reduced costs, improved productivity, and operational efficiency can reasonably be expected to be attained.

6.3.1.1 *RFID Technology Evidence Handling and Property Control*

Law enforcement is moving slowly beyond simple bar coding identification of evidence and equipment. RFID technology can be utilized as a more effective means of recording, locating, and tracking both evidence and property. RFID tag systems are able to record the who, what, where, and when of each piece of evidence in police custody. The courts emphatically require police agencies to precisely track and account for crime evidence from its initial seizure by police to its introduction at time of trial. The police must know where evidence is at all times and monitor where its been within the chain-of-custody. This calls for a high-visibility system, precise data collection and an accurate tracking process that will ensure the continuing integrity of the crime evidence handling process.

Law enforcement agencies are now able to apply RFID item-level "smart shelf" capabilities (mentioned earlier) and RFID tag systems to secure evidence and automatically meet the important chain-of-custody requirement. These law enforcement RFID systems can also send notifications or sound alerts if tagged evidence is removed from its secure storage location or if an unauthorized person attempts to handle it. This level of security can also extend to the tracking and recording of other police assets and property, i.e., laptop computers, handguns, shotguns, and other valuable police equipment, as well as the association of the particular asset to specifically assigned personnel.

6.3.1.2 *RFID Use in Police Patrol*

In April 2004, a police department in India initiated a pilot project in two police stations using RFID to track police officers on the beat. The project involves embedding 45 RFID chips at specific points within the police station limits and a RFID reader that is carried by the police officer. The system enables the police department to dynamically monitor police officer movement on the beat and helps management chart out and alter the beats in tune with the requirements of the area. The pilot initiative may be gradually expanded to cover the whole city in later phases of the project.

6.3.2 Applying RFID Technology to Ensure Police Officer Safety

RFID technology combined with other technologies, can provide a measure of safety to police officers. Examples include further development of "smart guns" technology to include RFID features and monitoring a police officer's location while on patrol.

6.3.2.1 *RFID Technology and "Smart Guns"*

Within law enforcement there is a need to find a better way to protect police officers from their own firearms. A FBI review looked at how law enforcement officers were killed and found that one in six was shot to death by their own firearm, after being disarmed by a suspect. In addition, 113 firearms were stolen from police officers during the same period.

The National Institute of Justice (NIJ) subsequently funded a study to examine the problem of firearms being taken away from police officers, to identify the extent to which officers are assaulted and killed with their own firearm, and to identify the requirements officers would want in a "smart gun." A "smart gun" could be a seamless, transparent technology solution that will allow a firearm to only be fired by a recognized and authorized user.

The NIJ study was followed up with a series of studies to identify the various technologies that could be used in developing a reliable "smart gun." The focus of the studies was on various biometric systems such as fingerprint technologies, computer chips that could be programmed to recognize an individual's grip or other physical features, and electronic codes and keys. Several firearm manufacturers have taken varied technical approaches to solving the problem of developing a reliable and effective "smart gun."

In April 2004, U.S. chip manufacturer Applied Digital Solutions announced a partnership with gun manufacturer FN Manufacturing to produce an RFID-enabled "smart gun." The technical approach taken by these two companies would require a tiny RFID chip to be implanted in a police officer's hand that would match up with a scanning device inside the handgun. If they matched, a digital signal would unlock the trigger to enable firing.

"Smart gun" technology research initially grew out of a concern for police offer safety and has been underway for approximately 10 years. RFID technology is the latest important addition to the pool of technical approaches to creating a reliable "smart gun" that could be marketed to law enforcement agencies for the purpose of ensuring police officer safety. The concept of a "smart gun" is feasible but not yet fully developed for police force adoption.

6.3.2.2 *RFID Monitoring of Police While on Patrol* In another example of a RFID law enforcement application to promote officer safety, the Orlando, Florida, Police Department attempted to pilot test a combined GPS/RFID location tracking system which would let the central office track police officers' locations. The presumed objective of the pilot project was to promote officer safety. The system was met with firm resistance from the police officers' union in that they felt it was intrusive to be continually monitored in their day-to-day police work and the project was canceled.

As a side bar, many companies monitor employee e-mail and internet usage, and security cameras are now commonplace fixtures in office buildings. However, technologies such as GPS and RFID tags promise to take employee monitoring to an entirely new level. Today's tracking can record, display, and archive the exact location of any employee, both inside and outside the office, at any time, offering managers the unprecedented ability to monitor employee behavior. Whether this is appropriate, needed, or required in law enforcement remains to be seen, but RFID technology readily provides the technical means to accomplish this at relatively low cost.

Despite the potential benefits of RFID tracking capabilities, even in the name of a noble objective such as ensuring police officer safety, employees of all kinds may find tracking technologies to be ominously intrusive. RFID technology can be an intrusive technology. However, it is probably inevitable that its deployment will become more commonplace over time. Only a social backlash may slow the growth of employee monitoring but it is unlikely to stop it.

6.3.3 Applying RFID Technology as a Crime Fighting Tool

The potential of RFID technology as a crime fighting tool is just now beginning to emerge. When combined with other technologies, such as GPS and biometric identification, RFID can provide police agencies with new and powerful technological tools to solve crimes. To what extent the application of the technology will be able to be fully deployed here in the United States, however, remains to be seen. At some level of use, constitutional issues will surely arise. Threats to personal privacy and infringements to civil liberties posed by RFID use by police agencies, however, are beyond the scope of this book. Ultimately, these issues will be resolved in the courts and through legislation as circumstances surrounding deployment arise.

6.3.3.1 RFID Technology and Property Crime Although still in its infancy, RFID technology has particular applicability in assisting the police in solving and/or preventing several types of crimes. As a tracking and tracing device RFID can be especially useful in addressing property crimes. RFID systems are expected to assist police in identifying and recovering stolen merchandise, and hence be a powerful deterrent to thieves, not only by increasing the risk of being caught but also by making it more difficult to find purchasers for the stolen merchandise. In addition, RFID systems are also expected to provide evidence in a court of law, which can help to convict those responsible for selling stolen merchandise.

As a counterfeiting detecting device, RFID systems allow the introduction of an unobtrusive marking that would detect fake items or "knock-offs" quite easily. Forging or copying RFID tags is very difficult, so it is simply a matter of scanning the product with a RFID reader to detect a counterfeit product.

In 2000, the United Kingdom launched the "Chipping of Goods" Initiative to show how property crime could be reduced using RFID technology. This strategic government initiative, in partnership with several major manufacturers of consumer goods, was initiated to show the effectiveness of chipped goods in combating crime and to accelerate the wider uptake of RFID technology. The initiative was in response to the need to reduce the cost of property crime, relieve pressure on police resources, and to trace the ownership of stolen goods. Some types of products included in the initiative were small boats, laptop computers, wine and spirits, and some consumer disposable products.

The initiative was designed to address key requirements of the Home Office and the police in terms of:

- Knowing whether goods have been stolen
- Providing proof of ownership
- Providing an audit trail to show where goods have been and who was involved in handling them during their life cycle

Government funding was matched by project private sector partners to establish eight demonstration projects to show the effectiveness of chipped goods in combating property crime.

6.3.3.2   RFID Technology and Automobile License Plates   In the United Kingdom, a company is developing an active RFID-enabled license plate with embedded long range RFID tags. The system will allow for speed checking sensors and other mechanisms to identify the automobile in real time from up to 300 feet away. The system is expected to be used for compliance with road taxes, electronic payment, tracking, insurance, vehicle theft and associated crime, and traffic counting and modeling. The reader network, which includes fixed and portable readers, sends a unique identifier in real time to a central system where it is matched with the corresponding vehicle data such as registration number, owner details, make, model, color, and tax/insurance renewal data.

Several German and South African companies are also working on RFID-enabled license plates to deter automobile theft and provide detailed information on the automobile and the registered owner.

6.3.3.3   RFID Technology and Drivers' Licenses   Recently, hearings were held in Virginia to explore the idea of creating a smart driver's license that eventually would include a combination of RFID tags and biometric data, such as fingerprint or retinal scans. The Virginia General Assembly wanted to deter fake identity documents, make it much harder to use a stolen or forged license for identification, and make look-ups faster for police officers and other government officials. Virginia didn't pass any legislation on the RFID-enabled driver's license, and the chairman of the committee conducting the hearings stated, "I can't see us using RFID until we're comfortable we can without encroaching on individual privacy, and ensure it won't be used as a Big Brother technology by the government."

The Virginia hearings were prompted by the introduction of federal legislation in the House of Representatives, the Driver's License Modernization Act of 2002, which called for the states to comply with uniform "smart card" standards. This would make state driver's licenses into de facto identity cards that could be read at any location throughout the country. The RFID chips on a driver's license would, at a minimum, transmit all of the information on a driver's license. This proposed post-9/11 federal legislation eventually lapsed without a vote.

The major objections voiced by privacy advocates at the Virginia hearings and the federal legislation hearings were that RFID tags in a driver's license are remotely readable and allow authorities to easily track citizens nationwide, using a state's driver license. Another fear was that an RFID driver's license could easily lead to the development of a national identification system without actually creating a national ID card.

Active interest in RFID driver's licenses has waned since 9/11 but the American Association of Motor Vehicle Administrators continues to advocate uniform standards among the states for drivers' licenses and does not object to RFID tags and biometric features being incorporated into drivers' licenses, subject to legislative approval and federal funding for implementation.

6.4 RFID USE IN LAW ENFORCEMENT—LOOKING TO THE FUTURE

Conceivably and some time in the future, any RFID-enabled object found at a crime scene, from an empty soda can to a knife, could be traced through the supply chain to a retail merchant. If the object was purchased with a credit card or a customer loyalty card it could be traced back to the initial purchaser, providing the police the identity of a potential witness or suspect to the crime. RFID applications with this type of crime-solving potential will eventually be recognized as a "must have" technology with unlimited potential for improving law enforcement processes.

While there are several private sector RFID technology firms in the United States that specialize in developing law enforcement RFID applications, current demand for RFID technologies is not widespread in law enforcement. However, as the rate and pace of RFID technology development and deployment accelerates, it appears to be only a function of time when forward-looking law enforcement agencies will acknowledge the efficiencies to be gained in deploying RFID technology in new and novel ways and begin to leverage RFID into their administrative, operational, and crime fighting processes.

6.5 RFID TECHNOLOGY IN CORRECTIONS

6.5.1 Background and Evolution of RFID Technology in Corrections

RFID technology for corrections applications grew out of military research conducted in the 1980s. Motorola Corporation initially developed RFID technology to track soldiers on the battlefield, but the end of the Cold War and budget cuts at the time determined that these RFID systems were unlikely to be rapidly adopted by the military. Motorola then began looking for a way to commercialize its RFID technology research and development. Motorola decided that its RFID system was better suited for prison operations, if it could be miniaturized. Motorola subsequently hired a former State Department of Corrections administrator to look at ways of using RFID to track and monitor inmates in a prison environment. Since prison management and prison operations were removed from Motorola's core competencies, it eventually licensed the RFID technology to Alanco Technologies, Inc., of Scottsdale, Arizona.

RFID technology applications for corrections evolved in much the same manner as they evolved in the commercial sector. Initially, barcodes were employed merely to replace or speed up the collection of data, such as, replacing logbooks, paper passes for inmate movements, cell checks, or the issuance of keys. At the next level, barcode technology became a warning mechanism to alert prison management if an inmate was late arriving and checking in at a location from his last destination or when a cell hadn't been checked at the required time.

Today, advanced RFID systems in corrections allow continuous inmate tracking to prevent escape, reduce violence, and continuously monitor and record the location of inmates and guards within the prison.

6.5.2 A RFID Technology Case Study in Corrections, Alanco Technologies, Inc.

In general, prisons introduce technology into their operations to produce cost-savings, particularly for labor-intensive tasks, such as prison guard services. Alanco Technologies, Inc., of Scottsdale, Arizona, believed they could generate substantial cost savings in prison operations through the use of their RFID technology and entered the prison security market in 2002, in part to eliminate the cost of continually conducting physical head counts, to reduce overall operating costs for the prison system, and to create an overall safer prison environment. Alanco developed its TSI PRISM RFID tracking system to address these prison operational needs.

The Alanco RFID tracking system consists of five primary components: a tamper detecting industrial-size wristwatch RF transmitter for inmates, a belt-mounted transmitter worn by the officer staff, a strategically placed array of receiving antennae, a computer system and proprietary application software. The system's software simultaneously processes multiple and unique radio signals received every two seconds from the prisoner's wrist and the guard's belt transmitters to pinpoint their location and track and record in real time as they move about the facility. Entry into restricted areas or attempts to remove the transmitter device signals an alarm to the monitoring computer. The guard's transmitter can also signal an alarm by manual activation of an emergency button, or automatically, if the guard is knocked down or the transmitter is forcibly removed from his belt. The system automatically conducts an electronic head count every two seconds.

The system provides real-time individual identification and tracking with its array of database and software applications. The system automatically records all tracking data over a prescribed period in a permanently archived database for accurate post-incident reporting and future reference. A host of management reporting tools are also available with the system that include medicine and meal distribution, adherence to time schedules, restricted area management, and specific location, arrival, and departure information.

6.5.3 Validation of Alanco's RFID System in a Prison

In late 1999, the first operational TSI PRISM system was installed at a minimum-security prison in Calipatria, California. By August 2002, the system successfully completed a comprehensive, 90-day testing program conducted by the California Department of Corrections (CDC).

During the trial period, Calipatria had several prisoner disturbances. After the guards regained control of the facility the system was quickly able to identify the prisoners involved in the disturbance and officials were able to take appropriate disciplinary action.

The RFID system also assisted in the recapture of a prisoner who escaped. A prisoner cut his wristband, which signaled an alert. A guard was sent to investigate and the prisoner was quickly recaptured, before he was a mile from the facility. Prior to the installation of the RFID system, two earlier escapes were not discovered until the next scheduled inmate head count, several hours after the inmate left the facility.

Typically in a prison, when there is a fight between two inmates or a stabbing, no one talks for fear of reprisals. Guards normally have to lock down the facility to conduct an extensive investigation. With the RFID system in place, a data review reveals the identification of the other inmates who were around the victim at the time of the assault. This enables the guard staff to interview only those around the victim at the time of the assault rather than a large segment of the prison population. As a consequence, the RFID system tends to reduce inmate violence and property damage in the prison because the system is able to show a particular inmate in a particular location at a particular time and the investigation can focus on these inmates.

The technology evaluation process for the Alanco RFID system at Calipatria included a 90-day evaluation report by CDC. The testing report included the following highlights:

- The TSI PRISM system aided in the early detection of an escape attempt, resulting in the inmate's capture within one-and-a-half hours
- The system accurately determined the identify of an inmate assault
- The system successfully resisted inmate attempts to tamper or otherwise defeat the system
- The system provided a continuous inmate headcount at two second intervals, proved effective in reducing staff time required to complete headcounts, and readily identified officers and their specific locations whenever a duress alarm was initiated

Based, in part, on the CDC evaluation report of the Alanco RFID system at Calipatria, the State of Michigan installed it in a high-security juvenile detention facility. The RFID system was adopted to protect the staff from inmates that claimed they were being assaulted by the guards. There had been numerous abuse complaints by inmates and the investigation and legal costs to resolve the complaints were mounting. Michigan focused on the Alanco RFID system in view of information contained in the CDC evaluation report that reported that, after using the system for two years, incidents of inmate violence had declined by 65%. The system was eventually expanded in Michigan to include other correctional facilities.

In October 2002, Alanco commenced installation of its RFID system at a medium security in an Illinois prison facility which was spread over 25 acres.

In August 2004, the Ohio Department of Rehabilitation and Correction (ODRC) approved a $415,000 contract with Alanco for a pilot RFID system installation project at the Ross Correctional Facility in Chillicothe, Ohio. The contract is a precursor for potential system-wide RFID installation throughout the Ohio Prison system's 33 separate facilities and its 44,000 prisoners.

6.5.4 Implanting RFID Chips in Prison Inmates

With the recent FDA approval of the human implantable VeriChip as a device that can be used for "security, financial, and personal identification/safety applications" (discussed earlier in this report), it is only a question of time before the "chipping" of prison inmates will be contemplated as a viable and effective RFID application to improving prison management and administration. In make corrections management and prison operations more secure, accountable, and efficient, the possibilities for inmate RFID implants are endless.

Consider the following. Most inmate record systems are intended to gather and provide easy access to information about inmates and their behavior within the correctional facility. Additionally, automating routine prison operations has always been a goal of prison administrators to lowering costs and improving safety.

Through inmate RFID implants, access to an inmate's electronic record could readily be available by way of the inmate's individualized RFID chip. For example, admission and release records, schedules and movements, legal documents, sentence administration, classification, offenses and custody, gang affiliation, property and clothing records, visitors, trust accounts, commissary, billing of services, medical information, and transportation schedules could be conveniently stored in a database and be accessed through the inmate's RFID chip, without the possibility of inmate misidentification or mistake.

As to prison operations, an inmate's RFID implant has the potential of automating many of the daily, yet very important, prison functions. For example, movement of inmates and visitors within the facility can be tracked and monitored through the implanted chip and a historical record of each movement could be maintained; alerts could be initiated when an inmate does not arrive on time at a designated location; queries could be initiated to locate each inmate or a list of all inmates at a particular location through the implanted chip; up to the minute status of inmate headcounts and cell checks could be maintained and immediately reconciled to identify missing inmates or staff; and inmate commissary and laundry paperwork can be eliminated.

While there is great appeal to the chipping of prison inmates as an effective technological solution that can contribute to lowering costs, improving operational efficiency and safety, it remain a very controversial procedure and raises social and ethical issues. Suffice it say that it is only a short distance from wearing an external watch-like RFID-enabled bracelet to "wearing" a subdermal RFID—enabled implant device. However, implanting the technology in the human body versus externally wearing the technoloy does not appear to be a functionally equivalent process.

Finally, it should be noted that potential research into the effectiveness of RFID implants of prisoners may require compliance with special federal requirements, particularly agencies, companies, and institutions that receive federal funding. The Office of Human Research Protection within the Department of Health and Human Services provides leadership and guidance on human research subject protections and implements a program of compliance oversight for the protection of human subjects participating in research. Additionally, specific rules also apply when prison inmates are used as research subjects.

6.5.5 Electronic Monitoring in Corrections

Electronic monitoring is a broad term that encompasses a range of different types of technical personal surveillance. Each type of electronic monitoring has the potential to be used in different ways and depending on the technology used, electronic monitoring can provide a continuous indication of location so that the whereabouts, or the presence or absence of a person at a location can be checked at any time. One form of continuous monitoring may involve the offender's movements being tracked so that his movements are known at any given time. Others can be used to restrict people from specified areas or individuals. In such cases, any alert requires to be reinforced by prompt action by the monitoring service providers or the police in order to protect a potential victim or enforce court-ordered sanctions.

Electronic monitoring of offenders was first developed in the United States in the 1970s, but took off in the early 1980s when it was seen as a cost-effective way of reducing burgeoning prison population. The initial focus of electronic monitoring took the form of house arrest, where the offender was sentenced to remain in the house (except when fulfilling other conditions of his court order) and compliance was monitored by an electronic tag worn on the ankle. Offenders sentenced to house arrest were typically low-risk but otherwise likely to be imprisoned.

Different models of wireless electronic monitoring address different situations. Global positioning systems provide the means to track the movements of offenders via satellite. However, the expense and intrusiveness of tracking technology is inappropriate for an offender who poses a low level risk. Equally, the use of tracking in cases where the intention of the court is to restrict the offender primarily as a penalty, rather than as a public safety measure would also be questionable. In such cases, electronic checks on the offender's presence or absence at the location to which they are restricted is probably be adequate. It should be recognized, however, that electronic monitoring of this kind provides no information about the whereabouts of the offender when they are outside the range of the equipment. Conversely, the use of tracking for higher risk offenders fully depends on the electronic monitoring system being foolproof.

Currently, most states use wireless electronic monitoring in some form, including for home detention, probation, parole, juvenile detention, and bail. It is estimated that there are about 1,500 electronic monitoring programs that involve about 100,000 offenders in the United States.

6.5.6   Global Positioning Systems (GPS) in Corrections

An RFID system for corrections may appear to compete with a wireless GPS, but GPS cannot "see through concrete" and it is not a very effective option in high-security prison use. Accordingly, GPS is more effectively used in such community correctional settings as juvenile detention, domestic violence, pretrial release, conditional release, and the tracking of known sex offenders where the offender poses a public safety risk.

The estimated cost of operating prisons and jails in the United States is over $57 billion per year. However, GPS tracking and monitoring costs about one tenth the cost of incarceration. This has become a major consideration for expanding the use of GPS in corrections. For every offender who can be removed from prison and subjected to GPS tracking, a prison space is made available for detaining and controlling a violent offender. It is estimated that GPS electronic monitoring cost between $4.50 to $12.00 a day versus $60.00 to $100.00 per day for incarceration.

In a typical case, an individual subjected to GPS tracking and monitoring wears a removable personal tracking unit (PTU) and a non-removable wireless ankle cuff the size of a large wristwatch. The cuff communicates with the PTU to ensure it remains in close proximity. If communication with the cuff is lost, the PTU records a violation.

The PTU acquires its location from the Department of Defense's GPS satellites and can communicate that information to an internet-based database system. Using a web browser, authorities can access a detailed map to determine where the individual has been. If the individual was in a place he or she was prohibited, the GPS tracking system would capture that information. Some systems provide detailed online mapping (denoted by color trails) of an individual's travels during a specified period, with zoom-in capability on street-level maps.

Some systems can set up exclusion zones and geo-code areas for those who are territory restricted. Automatic alerts can notify authorities when exclusion zones have been entered by the individual.

States are continuously seeking ways and methods other than prison to keep tabs on violent offenders and GPS tracking enables authorities to keep closer tabs on offenders who may pose a significant danger to their communities. Society is now demanding that the more than 600,000 convicted sex offenders currently out in the public be tracked and monitored. They want these people watched continuously and cost effectively. Accordingly, numerous states have passed legislation mandating the GPS tracking of sex offenders and it is anticipated that the future use of GPS-based electronic monitoring systems will rapidly expand to address this public demand.

6.5.7   RFID Technology's Future in Corrections

It is no secret that correctional facilities in the United States have historically been hampered by overcrowding, high operational costs, and general under funding. These issues have recently become a catalyst for change and for many state legislatures to turn to technology and private management and operation of public prisons. Today, Corrections Corporation of America and Wackenhut Corrections Corporation, two leading private sector prison management companies, manage many state and local prisons and jail facilities throughout the country. These companies have been able to rapidly grow by putting forth a compelling value and cost saving proposition to the states and to corrections officials.

Similarly, corrections officials are also aware that by incorporating RFID technology into a facility's operation, it also offers a similar compelling value and cost saving proposition. RFID technology in a prison's operation radically changes the way prisons are managed and operate. RFID technology transforms routine manual tasks that require costly manpower to accomplish to simple electronic tasks that can be accomplished effectively and at minimum cost. Ultimately, RFID systems provide real value, promote operating savings, decrease violence, create a safer work environment for inmates and staff, and create a more effective and efficient prison system.

CHAPTER 7

RFID REGULATIONS
AND STANDARDS

7.1 GOVERNMENTAL RFID REGULATION

RFID is a radio communication technology, and as such it is subject to governmental regulation in most countries. Governmental regulation is required to coordinate the use of electromagnetic spectrum between competing uses, such as radio, television and mobile phone systems, as well as to protect the public's interest.

Governmental regulations are necessary to accomplish the following:

- Establish order on the airwaves through the allocation and licensing of electromagnetic spectrum to users. The use of the spectrum must be coordinated among the many applications competing for bandwidth, including RFID applications, or chaos will ensue, rendering the spectrum useless. In order to ensure an equitable division among its many users, governmental regulators license segments of the spectrum to individual operators. These licenses are very specific about the permissible uses for the spectrum. For example, some segments of the spectrum are licensed only to TV broadcasters; others are licensed only for satellite communications while still others are only for mobile phone operators.

- Establish best practices and safety guidelines. Regulations are required to protect the public's interests, as well as its health. For example, a

RFID-A Guide to Radio Frequency Identification, by V. Daniel Hunt, Albert Puglia, and Mike Puglia
Copyright © 2007 by Technology Research Corporation

regulatory agency might prevent one organization from holding too many TV licenses in the same market, in order to ensure a diversity of voices in the media. Regulations are also required to limit human exposure to electromagnetic radiation. This is most often accomplished through placing power limitations on radiators and setting rules on the placement of radiating antennas. For example, mobile phones are limited to 1 watt in the United States and the antennas on cellular telecommunications facilities must be placed a certain minimum distance from public thoroughfares. This type of regulation also applies to RFID interrogators.

- Establish maximum permissible interference guidelines. All users of the electromagnetic spectrum will interfere with one another to some extent. Regulations are required to set the upper limit on how much one radiator may interfere on another's band. In addition, processes are required to enforce licenses and to hold licensees to task. In the event that one licensee complains about interference from another, an investigative procedure must be in place to remedy the situation. This is important for RFID applications, since it is assumed that some day many different RFID systems could be operating in the same enclosed space (a shopping mall for instance).

7.2 WORLD REGULATORY BODIES

The major players in the RFID industry are found in the United States, Japan, and in several European nations. The regulating bodies in these countries have considerable influence over the direction RFID technology will take in the coming years.

- In the United States, the FCC regulates the electromagnetic spectrum.
- In Japan, the Ministry of Public Management, Home Affairs, Posts and Telecommunications (MPHPT) fulfills the role.
- In Europe, the situation is a bit complicated. Each of the European nations has its own regulatory body, however, most of them are concurrently united under two organizations, among which the responsibilities given to the FCC and the MPHPT are divided. Both of these European organizations are in one way or another tied to the European Conference of Postal and Telecommunications Administrations (CEPT).

The first of these organizations is the European Radiocommunications Office (ERO), which supports the European Communications Committee (ECC), which is formerly the European Radiocommunications Committee (ERC). The ECC in turn is the telecommunications regulation committee for CEPT, mentioned previously. Its main task is to develop telecommunications policies and to coordinate frequency and technical matters for its 46 member countries, to create a uniformity in standards across Europe. The ERO publishes and distributes ECC decisions and recommendations.[49]

The second of these organizations is the European Telecommunications Standards Institute (ETSI), which was created by CEPT to establish consensus-based telecommunications standards for its 55 member countries. ETSI has published several RFID standards and has been playing a much greater role in regulating RFID than the ERO.[50]

7.3 INDUSTRIAL-SCIENTIFIC-MEDICAL (ISM) BANDS

Most RFID systems are designed to operate in so-called Industrial-Scientific-Medical bands. ISM bands are special license-free bands that have been set aside by regulatory bodies across the world. Originally intended for non-commercial industrial, scientific, and medical uses, they are now being used for a variety of commercial applications, such as wireless LANs and Bluetooth, in addition to RFID. As a result, By using ISM bands RFID system operators can skirt the licensing process that other wireless telecommunications operators are forced to undergo. ISM bands are not unregulated however. There are still many ISM rules regarding use of the band, limits on radiated power, and tolerance of interference.

7.4 SPECTRUM ALLOCATIONS FOR RFID

As mentioned previously, there are four major RFID bands: LF, HF, UHF, and microwave. Spectrum allocations within these bands are not the same the world over, however. Significant differences do exist between the United States, Europe, Japan, and China. There is much more uniformity at the lower LF and HF than at the higher UHF and microwave frequencies:

- Low Frequencies (LF)—125–134 kHz is available for use in the United States, Europe, and Japan. RFID shares this band with aeronautical and marine navigational uses.
- High Frequency (HF)—13.56 MHz is also available for use in the United States, Europe, and Japan at very similar power levels.
- Ultra High Frequency (UHF)[51]—Attention is focused largely on the UHF band at present, as most emerging RFID applications will use this band. There are considerable differences between regulations in the United States, Europe, and Japan.

---

[49] *About RFID Regulations*, Impinj, www.impinj.com/page.cfm?ID=aboutRFIDRegulations.
[50] *About RFID Regulations*, Impinj, www.impinj.com/page.cfm?ID=aboutRFIDRegulations.
[51] *About RFID Regulations*, Impinj, www.impinj.com/page.cfm?ID=aboutRFIDRegulations.

• Microwave—An ISM band at 2.45 GHz is available in most regions, though the exact details vary. Four watts transmitted power is permitted in most places, though only 1 W is permitted in Japan. This ISM band is also used by wireless LAN and Bluetooth applications.

In the *United States*, RFID devices operating at UHF frequencies are allowed to use ISM bands under certain conditions. The UHF ISM bands are located between 888–889 MHz and 902–928 MHz in the United States. RFID interrogators are allowed to operate at 1 watt (W) transmitted power, or 4 W with a directional antenna if the interrogator employs frequency hopping.

In *Europe*, RFID regulations limit the transmitted power, channel bandwidth, and duty cycle of UHF interrogators, in comparison to the United States. At present, interrogators are limited to 500 mW transmitted power, though there are plans to increase that to 2 W. The ERO has specified a UHF band between 868 and 870 MHz for RFID use. The U.S. ISM bands are not available for RFID use because GSM mobile phone systems occupy those bands in Europe.

In Japan, there are no UHF frequencies for which RFID operation is permitted. A band has been recently opened up between 950 and 956 MHz for experimentation. Australia has designated a band between 918 and 926 MHz for RFID use, with a 1 W transmitted power limit.

7.5 INDUSTRIAL RFID STANDARDS

To date, the RFID industry has been driven by diverse, vertical application areas. Most RFID systems on the market are proprietary systems as a result. This had been recognized as a barrier to widespread RFID adoption and industry growth. Emerging applications will require the inter-operation of RFID products from different suppliers, as well as the inter-operation between RFID systems in different countries and regions. For this reason, a worldwide effort is being made to standardize RFID systems.

The purpose of RFID standards is to create a degree of product uniformity in the RFID industry, in order to enhance the efficiency of RFID systems, which ultimately will make RFID more cost effective and lead to industry growth. They give industry participants a common platform from which to move forward. Whereas regulations are set by government entities, RFID standards are set by standards bodies.

There are many standards bodies around the world addressing this issue. They include:

• International Organizations
  International Organization for Standardization (ISO)
  International Electro-technical Commission (IEC)

International Telecommunications Union (ITU)

EPCglobal

· Regional Organizations

European Conference of Postal and Telecommunications Administrations (CEPT)

European Telecommunications Standards Institutes (ETSI)

· National Organizations

British Standards Institute (BSI)

American National Standards Institute (ANSI)

This chapter will focus on ISO and EPCglobal standards, as they are seen to have the greatest influence on the RFID industry at present.

7.6 INTERNATIONAL STANDARDS ORGANIZATION (ISO)[52]

ISO and IEC have formed a joint subcommittee, called the ISO/IEC JTC1. The JTC1 is divided into subcommittees, some of which address the standardization of RFID technologies. In 2006 ISO adopted the key EPCglobal RFID Standards.

7.6.1 Standards for RFID Animal Tracking

Very few standards exist for LF RFID systems, due to the fact that most LF RFID applications are in closed-loop controlled environments and there is little need for inter-operability between systems. Animal tracking systems, however, which use LF, have required some standardization. ISO has developed two standards for this purpose, ISO 11784 and ISO 11785, and they have met with some, though limited, industry acceptance:

· ISO 11784—RFID of animals—code structure. This standard defines the code structure for animal tags. Animals can be identified by country code and a unique national ID.
· ISO 11785—RFID of animals—technical concepts. This standard defines the technical parameters of tag/interrogator communication.

7.6.2 Standards for RFID Identification Cards and Related Devices

The 17th subcommittee, 8th working group (SC17-WG8) of JTC1 was formed to address the standardization of RFID identification cards and related

---

[52]*White Paper: Demystifying RFID: Principles and Practicalities*, Steve Hodges and Mark Harrison, Auto-ID Centre, 2003.

devices. The work began in 1995 and three standards were published in 2000: ISO 10536, ISO 14443, and ISO 15693. These are the most widely used and accepted RFID standards to date, however, they only pertain to HF RFID systems.

- ISO 10536—Identification cards and contactless integrated circuit cards. This standard describes the parameters for proximity coupling smart identification cards, with a read range of 7–15 cm, using 13.56 MHz. There are four parts to the standard:

  Part 1: Physical characteristics
  Part 2: Dimensions and locations of coupling areas
  Part 3: Electronic signals and reset procedures
  Part 4: Answer to reset and transmission protocols
- ISO 14443—Identification cards and proximity integrated circuit cards. This standard also describes the parameters for proximity coupling smart identification cards, with a read range of 7–15 cm, using 13.56 MHz. There are four parts to the standard:

  Part 1: Physical characteristics
  Part 2: Radio frequency power and signal Interface
  Part 3: Initialization and anti-collision
  Part 4: Transmission protocols
- ISO 15693—Contactless integrated circuit cards and vicinity cards. This standard describes the parameters for vicinity coupling smart identification cards, with a read range of up to 1 m, using 13.56 MHz. There are four parts to the standard:

  Part 1: Physical characteristics
  Part 2: Air interface and initialization
  Part 3: Protocols
  Part 4: Extended command set and security functions

7.6.3 Standards for RFID AIDC and Item Management Technologies

The 31st subcommittee, 4th working group (SC31-WG4) of JTC1 was formed to address the standardization of RFID Automatic Identification and Data Capture (AIDC) technologies, as well as item management technologies. Some of these standards have been published as recently as 2004, and they include ISO 15961, ISO 15962, ISO 15963, ISO 18000, and ISO 18001. A major standardization step by ISO was the adoption/support of the EPCglobal RFID standards. These are the standards that apply to supply chain and asset management systems. They address the full range of RFID frequencies, LF to microwave, and their adoption is viewed to be very important to the promotion of RFID technology.

TABLE 7-1   ISO/IEC 18000 Parts

Part 1	Generic Parameters for Air Interface Communication for Globally Accepted Frequencies
Part 2	Parameters for Air Interface Communication below 125 KHz
Part 3	Parameters for Air Interface Communication at 13.56 MHz
Part 4	Parameters for Air Interface Communication at 2.45 GHz
Part 5	Parameters for Air Interface Communication at 5.8 GHz
Part 6	Parameters for Air Interface Communication at 860–930 MHz
Part 7	Parameters for .Air Interface Communication at 433 MHz

Source: Auto-ID Center/EPCglobal.

- ISO 15961—RFID for item management—data protocol and application interface specification. This standard defines the functional commands and syntax features of item management systems, for example, RFID tag-types, data storage formats, compression schemes, etc. These protocol specifications are independent of transmission media and air interface protocols. Its companion standard is ISO 15962, which provides the overall protocol for data handling.

- ISO 15962—RFID for item management—data protocol, encoding rules and logical memory functions specification. This standard specifies the interface procedures used to exchange information in an RFID system for item management.

- ISO 15963—RFID for item management—specification for the unique identification of RF tags. This standard specifies the numbering system, the registration procedure and the use of uniquely identifiable RFID tags.

- ISO 18000—RFID air interface standards. The ISO 18000 standard provides a framework for defining common communications protocols for international use of RFID. Where possible, it also specifies the use of the same protocols for different frequency bands (LF, HF, UHF, microwave) to minimize the problems of migration and to make RFID platforms similar across the spectrum of bands. The ISO 18000 specifications are divided into seven parts, as shown in Table 7-1.

- ISO 18001—RFID for item management—application requirements profiles (ARP). Published in 2004, this standard address information technology standards in RFID systems.

7.7   EPCGLOBAL

Wal-Mart and DoD both specified the use of EPCglobal RFID technology standards in their RFID mandates. Other major retailers, such as Target and Metro AG, the leading retailer in Germany, have also adopted the standards developed by EPCglobal. As a result, the EPCglobal standards appear to be the standards of choice for retailing and supply chain management applications, and it is believed that their standards will have a great influence over the direction the technology and industry ultimately takes. Note in all reference cases in this book, EPCglobal is a recognized trademark for EPCglobal.

7.7.1 History of EPCglobal

EPCglobal traces its beginnings to an academic research center based at MIT called the Auto-ID Center. The Auto-ID Center, founded in 1999, is a partnership between over 100 global companies and five universities around the world: MIT in the United States, Cambridge University in the UK, the University of Adelaide in Australia, Keio University in Japan, and St. Gallens University in Switzerland. Together, they are working to build the technology standards and system components necessary to apply RFID technology to inventory and supply chain management.

The Auto-ID Center's goal is to create an "internet of things," as opposed to an internet of computers. In other words, they are building a global infrastructure, "a layer on top of the internet," that will make it possible for computers all over the world to uniquely identify tagged objects instantly. They call this "internet of things" the Electronic Product Code Network. They are designing the critical elements of this network: the Electronic Product Code (EPC), specifications for cheap RFID tags and cost-effective interrogators, an Object Naming Service (ONS), a Product Markup Language (PML), and Savant software application technology.

In 2003, EPCglobal was created through a joint venture between the Uniform Code Council (UCC), makers of the UPC symbol, and EAN International. It is a non-profit organization entrusted by the RFID industry to support and establish standards for the EPC Network. Its goal is to promote the adoption of the EPC Network standard. The administrative work of the Auto-ID Center was assumed by EPCglobal when it was created. The Auto-ID Center is now referred to as Auto-ID Labs and constitutes the research wing of EPCglobal. Auto-ID Labs will continue to conduct research in support of the Electronic Product Code Network.

7.7.2 The EPC Network

The EPC Network is composed of four basic components: an object tagged with an EPC label, a computer system running Savant, an Object Name Service (ONS) server, and a Product Markup Language (PML) server. The Savant computer, ONS server, and PML server are most likely connected through the internet and very far apart from one another.

To illustrate how the EPC Network operates, see Figure 7-1. An object, such as a can of soda, is tagged with an EPC label. The EPC label stores a

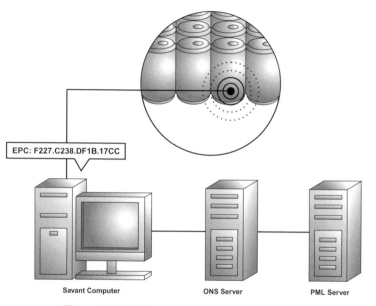

EPC: F227.C238.DF1B.17CC

Savant Computer   ONS Server   PML Server

Figure 7-1 EPC Network. Source: EPCglobal.

code number, a unique identifier, that indicates what company manufactured the can of soda, as well as the serial number for that particular can.

A Savant computer, which is essentially a network of interrogators and a host running application or software, reads the EPC label on the package. This can occur anywhere in the product chain. Multiple Savant computers and readers could be installed at the factory, at distribution centers, at warehouses, or at retail locations. Let's assume this Savant computer and interrogator are installed at a retail location.

Once the Savant computer has read the EPC label of the can, it forwards the code number to an ONS server, which is akin to a reverse phone book. The ONS server is able to take the EPC number and produce the name and address of the company that manufactured the can of soda. It then forwards that name and address back to the Savant computer.

The Savant computer can use the name and the address of the company to contact that company's PML server directly. Just as all companies have a website and a web server, in the EPC network, all companies will have a PML site and a PML server. Suppose the manufacturer of this can of soda is Pepsi. The Savant computer at the retail location would contact Pepsi's PML server with the unique serial number of the can of soda. The Pepsi PML server would contain all kinds of information about that particular can, such as the date and location of production, whether or not the product has been recalled, what points it has passed through in the distribution chain, etc. The Savant computer could check on this information to make sure the can of soda is suitable for sale. Furthermore, if it were the last can of Pepsi on the shelf, the Savant computer could order more product. All this could be done with little or no human interaction. This, in short, is how EPC Networks are designed to operate.

On a side note, now that the basic operation of EPC networks has been explained, it might be useful to draw some parallels between the internet and the EPC Network. Whereas PML servers are akin to web servers, ONS servers are similar to internet Domain Name Servers (DNS). On the internet, every time a user clicks on a web link, a DNS is contacted to translate the web link, i.e., www.yahoo.com, into an IP number, i.e., 123.456.789.10, which is a web server's numerical address. The actual website can not be contacted until the user's computer has this information. The same applies to EPC networks. Furthermore, whereas there are thousands of web servers around the world, there are relatively fewer DNS and they are centrally controlled, ultimately by the U.S. government. In other words, there can only be one phone book for the internet. The same would ostensibly be true for the EPC Network. Whereas every company would have its own PML server, there would be relatively fewer ONS servers under some sort of central control.

7.7.3 EPC Standards

EPCglobal is developing standards and specifications for the following EPC Network components:

- EPC Tag Data Specifications
- Communications interfaces for HF and UHF systems
- Reader Protocol
- Savant
- Object Name Service (ONS)
- Physical Markup Language (PML)

7.7.4 EPC Tag Data Specifications

The EPC is similar to a Uniform Product Code (UPC), the bar code standard. EPCglobal is not only trying to establish a migration path for companies to move from bar codes to RFID, however. They are creating a larger umbrella group called Global Trade Identification Numbers (GTIN), under which both UPC and EPC symbols will fall. It is not clear yet that these proposals will be accepted by industry, however, EPCglobal does have the support of UCC, maker of the UPC symbol and one of its parent companies.

An EPC number is made up of several parts. There is a header and three sections for data, as shown in Figure 7-2. The EPC is sometimes referred to as a "license plate" code.

The header identifies the EPC version or type number, in this case Version 1. The second part identifies the "manager," which is most likely the product manufacturer, for example, Pepsi. The third part identifies the object class, or the exact type of product, for example, Diet Pepsi, 12 oz. can, U.S. version. The final section identifies the serial number for the can of Pepsi the tag is attached to.

There are two versions of tags at present: one version has a memory that is 64 bits long and the other, shown in Figure 7-2, is 96 bits long. Larger tag memories may be possible in the future, but 96 bits seems sufficient to meet the world's current needs. A 96-bit tag, apportioned as shown in Figure 7-2, with 28 bits for the manager number, 24 bits for the object class, and 36 bits for the serial number, will uniquely identify 268 million companies, each of which could have up to 16 million different products and 68 million unique serial numbers for each product; more than enough to cover all the world's manufactured goods for many years to come. The smaller 64-bit tags were designed to fulfill a short-term need of the industry, as they are cheaper to manufacture than the 96-bit tags and will help to keep the cost of initial implementation down.

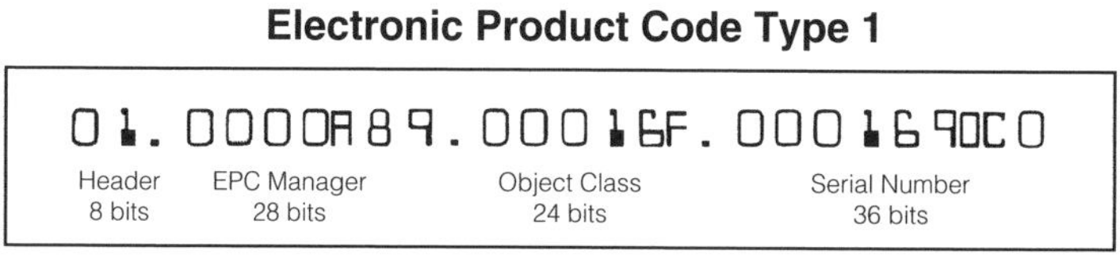

Electronic Product Code Type 1

| Header | EPC Manager | Object Class | Serial Number |
| 8 bits | 28 bits | 24 bits | 36 bits |

Figure 7-2   EPC Type 1 Tag Numbering System. Source: EPCglobal.

In addition to two versions of tags, there are several classes of EPC tags. The characteristics of these tags are summarized in Table 7-2.

TABLE 7-2   Classes of EPC Tags

EPC Class	Definition	Programming
Class 0	Read-Only Passive Tag	Programmed During the Semiconductor Manufacturing Process
Class 1	Write-Once-Read-Many Passive Tags	Programmed Once by the End User
Class 2	Re-Writable Passive Tags	Can be Reprogrammed Many Times
Class 3	Semi-Passive Tags	Can be Reprogrammed Many Times
Class 4	Active Tags	Can be Reprogrammed Many Times

Source: Impinj.

7.7.5 Published Specifications

The specifications for other EPC Network System components are summarized in Table 7-3.

TABLE 7-3 Published EPC Standards

EPC Tag Data Standards	Specific Encoding Schemes for a Serialized Version of the EAN.UCC Global Trade Item Number (GTIN®®), the EAN.UCC Serial Shipping Container Code (SSCC®®), the EAN.UCC Global Location Number (GLN®®), the EAN. UCC Global Returnable Asset Identifier (GRAI®®), the EAN.UCC Global Individual Asset Identifier (GIAI®®), and General Identifier (GID).
UHF Class 0 Specifications	Communications Interface and Protocol for 900 MHZ Class 0
UHF Class 1 Specifications	Communications Interface and Protocol for 860–930 MHZ Class 1
HF Class 1 Specifications	Communications Interface and Protocol for 13.56 MHZ Class 1
Radar Protocol	Communications Messaging and Protocol Between Tag Readers and EPC Compliant Software Applications
Savant Specifications	Specifications for Services Savant Performs for Application Requests Within the EPCglobal Network
Object Name Service Specifications	Specifications for How the ONS is Used to Retrieve Information Associated with a Electronic Product Code (EPC)
Physical Markup Language Core Specification	Specifications for a Common Vocabulary Set to Be Used Within the EPC Global Network to Provide a Standarized Format for Data Captured by Readers

Source: EPCglobal.

7.7.6 Future EPC Standards

It should be noted that EPC tags do not hold much more data than UPC bar code symbols and can not be written to. As such, some argue that EPC is not taking full advantage of the benefits offered by RFID technology. EPCglobal argues that this design lends itself to a cheaper EPC tag.

There is now a second generation of EPC labels, sometimes referred to as Gen 2. (The current set of published standards, described above, is referred to as EPC Generation 1.) It is intended that the current class structure will remain in place in Gen 2, however, tag functionality will be increased. This redesign is motivated in part by the wishes of Wal-Mart, DoD, and other retailers to have more flexible tag data structures with user-defined, re-writable sections of memory, as opposed to just static product numbers or "license plates." The Gen 2 standards has been published an adopted by the ISO standards organization. They largely pertain to UHF implementations.

7.8 THE WAL-MART AND DOD MANDATES AND EPC

The initial Wal-Mart RFID mandate specified the use of EPC Class 0 and Class 1, 96-bit Gen 1 tags. Sixty-four-bit tags are not being supported.

Wal-Mart has stated that they are now adopting the implementation of the Gen 2 EPC standard.

The DoD has outlined two options for product manufacturers in its RFID mandates. Manufacturers may either use a DoD data construct or one of several EPC label standards. It will accept both Class 0 and Class 1, 64-bit and 96-bit Gen 1 EPC tags for the short term. Once the EPC Gen 2 tags and readers are commercially available, DoD will also phase out the use of Gen 1.

CHAPTER 8

ISSUES SURROUNDING THE DEPLOYMENT OF RFID TECHNOLOGY

8.1 INTRODUCTION

A number of important implementation issues need to be addressed before there is widespread adoption of RFID technology. The most important impediments in the development of RFID technology are:

- Resolving consumer privacy issues
- Overcoming the costs of developing and deploying RFID technology
- A lack of global standards (China, etc.) and regulations
- Technological immaturity and integration with legacy systems
- Lack of robustness
- Lack of knowledge and experience, end-user confusion and scepticism
- Resolving ethical concerns
- Data management

8.2 PRIVACY ISSUES IN APPLYING RFID TECHNOLOGY

In the U.S. consumer-driven economy, personal privacy is protected by a complex and interrelated structural body of legal rights and regulations, consumer protections, and industry and business policy safeguards. However, to

RFID-A Guide to Radio Frequency Identification, by V. Daniel Hunt, Albert Puglia, and Mike Puglia
Copyright © 2007 by Technology Research Corporation

privacy advocates, RFID technology has the potential of impacting these personal privacy protections.

As with other emerging technologies, RFID has created never-seen-before personal privacy issues by making it possible to capture personal information about a consumer that was previously difficult or impossible to obtain. These privacy issues may ultimately act to deter or limit the full realization of the economic potential that RFID technology holds for consumers, businesses and the economy.

According to privacy advocates, RFID technology is capable of developing a detailed personal profile of a consumer based on a record of interconnected retail transactions that may link name, address, product purchased, or service used with other personal information. RFID technology offers the ability to particularize and monitor consumer activity and transactions. This has given rise to fears of "consumer profiling." Many privacy advocates are thus opposed to the full implementation of item-level RFID systems without additional privacy protections.

Those who are concerned about the ubiquitous use of RFID technology fear that it will undermine consumer privacy, reduce or eliminate purchasing anonymity, and threaten civil liberties. Privacy advocates such as the Electronic Frontier Foundation (EFF), the Electronic Privacy Information Center (EPIC), and Consumers Against Supermarket Privacy Invasion and Numbering (CASPIAN) have called attention to these privacy concerns. These groups have pressed for boycotts of companies utilizing RFID, called for legislation regulating RFID use, and raised the debate about RFID into the public domain. They worry that, if left to spread by its own economic momentum, RFID will become widely entrenched "without giving the public the necessary time to consider whether and to what extent they want RFID to proliferate.[53]

8.2.1   "Smart Shelf," RFID Tags, and the Rise of Privacy Concerns

In early 2003, Wal-Mart, Proctor & Gamble, Gillette, and U.K.-based supermarket chain Tesco teamed up to conduct various trials of an RFID-enabled "smart shelf" system in retail stores. The smart shelf system was designed to scan the contents of the store shelf and alert store employees via computer when supplies were running low or when theft was detected. The trials were widely seen as the initial step by retailers to push RFID technology from supply and warehouse management applications to in-store consumer use that could prevent shoplifting and speed shoppers through automated checkout lines. Soon after Wal-Mart first discussed its smart shelf evaluation, privacy advocates began to raise concerns about the technology. The primary ques-

---

[53] Harry Surden, *Unbundling the Privacy Debate: RFID, Privacy and Emerging Technologies, Radio Frequency Identification—RFID Privacy Law and Theory*, Stanford University, 2004 (http://www.stanford.edu/~hsurden/RFID Privacy Law.htm).

tions raised were: Would retailers and manufacturers be able to monitor products after consumers purchased them? Could the technology be misused by hackers and criminals or be exploited for government surveillance?

By July 2003, Wal-Mart announced that it was deferring its smart shelf evaluation. The benchmark evaluation had planned to use RFID technology to monitor how many razor blades were on the shelf in a Brockton, Massachusetts, Wal-Mart.

While Wal-Mart declined to explain why it deferred the smart shelf test, it reasserted its commitment to RFID. It sent a letter to suppliers telling them it would still require its top 100 suppliers to put RFID tags on all pallets and cases shipped to its distribution centers and stores beginning in 2005. Wal-Mart reiterated its mandate to utilizing RFID technology in its warehouse and supply chain operations.

In spring 2003, Benetton Clothing Co. announced that it was planning to evaluate embedded RFID chips that had been placed in the labels of every new garment bearing its Sisley brand in order to track the garment through the supply chain. This set off a storm of controversy among privacy advocates who called for a boycott of all Benetton clothing. Privacy advocates claimed that the embedded chips could be used to link the consumer's name and credit card information to the serial number in the garment, in essence "registering" the garment to the consumer. Further, any time the consumer went near an RFID reader device, the garment would identify the consumer, without his or her knowledge or permission. The controversial Benetton plan to RFID tag garments energized the privacy advocacy community to action.

Benetton moved quickly to downplay the RFID evaluation and agreed to remove the tags for retail consumers, upon request. While indicating that it planned to go forward with its RFID tag evaluation, Benetton quietly reexamined its initial position on RFID tags and has made no decisions concerning implementation. It is "studying the business case for implementing the technology and would consider the potential implications relating to individual security before firming up its RFID plans."

The Gillette Company, in conjunction with Tesco, Britain's biggest supermarket group, began an evaluation of a smart shelf that contained Gillette's razors in February 2003. The smart shelf was designed to contain packages of the Gillette Mach 3 razor, each package containing an RFID tag. The shelf contained a reader and a controversial small CCTV camera. According to reports, each time a razor package was removed from the shelf the RFID tag triggered the camera and a picture of the consumer was taken. The system also recorded an image of the consumer at the cash register when the razor was paid for. Consumers that were tape recorded at the shelf but not at the cash register potentially could have been suspected of shoplifting.

A group appeared outside the store in June 2003, protesting that the smart shelf was secretly monitoring customers. Tesco claimed it ended the trial as originally scheduled and was not affected by the protest. Gillette, on the other hand, said that the smart shelf evaluation should not have been used in conjunction with camera monitoring and Gillette and its retailers were shifting their attention to deploying RFID technology to track bulk shipments within warehouses and the company's supply chain.

In November 2003, it was revealed in the press that Proctor & Gamble conducted consumer testing of Max Factor Lipfinity lipstick at a Wal-Mart store in Broken Arrow, Oklahoma. The lipstick had RFID tags attached that allowed the inventory to be tracked leaving the shelf. The test also utilized a video camera that allowed researchers at Proctor & Gamble headquarters.

The privacy advocates called for mandatory labeling of products with RFID tags. Proctor & Gamble stated there was a sign near the lipstick shelf alerting customers that closed circuit televisions and electronic merchandise security systems were in place in the store. Further, Proctor & Gamble insisted the RFID system could only track lipstick leaving the shelf. Once the product was taken away, it was out of range of the RFID reader.

While limited consumer testing of item-level tagging RFID technology is continuing in Europe, it appears that it has been delayed in the United States, based on the privacy issues raised in the early trials. For example, Wal-Mart the primary commercial driver of RFID technology development in the United States has announced that it will focus on developing and applying RFID technology to its supply chain management and inventory control practices in the near term.

8.2.2   Consumer Privacy Concerns of Privacy Advocates

Judging by the excitement and controversy generated by the news stories about RFID item-level testing, it has become apparent that the closer RFID technology gets to the actual consumer, the hotter the privacy issues become. Privacy groups claim that numerous privacy issues need to be addressed prior to large-scale implementation of item-level tagging and the risk of abuse must be reduced.

According to privacy advocates, RFID technology, if used improperly, could jeopardize consumer privacy, reduces or eliminates purchasing anonymity, and threatens civil liberties. The principal consumer privacy and civil liberties organizations issued a position statement on "item-level RFID technology."[54] The following is a brief summary of their position for mitigating the risks to consumer privacy when RFID technology is applied to item-level tagging:

- Hidden Placement of Tags—RFID tags can be embedded into/onto objects and documents without the knowledge of the individual who obtains those items.

---

[54] *Position Statement on the Use of RFID on Consumer Products*, Issued by Consumers Against Supermarket Privacy Invasion and Numbering (CASPIAN), Privacy Rights Clearinghouse, November 14, 2003.

- Unique Identifiers for All Objects Worldwide—The use of unique product identification codes could lead to the creation of a global item registration system in which every physical object is identified and linked to its purchaser or owner at the point of sale or transfer.
- Massive Data Aggregation—RFID deployment requires the creation of massive databases containing unique tag data. These records could be linked with personal identifying data, especially as computer technology expands.
- Hidden Readers—Tags can be read from a distance and can be incorporated invisibly into nearly any environment where human beings or items congregate, making it impossible for a consumer to know when or if he or she is being "scanned."
- Individual Tracking and Profiling—If personal identify were linked with unique RFID tag numbers, individuals could be profiled and tracked without their knowledge or consent.

The privacy advocates go on to recommend a three part framework of rights and responsibilities to mitigate the consequences of RFID technology. The framework emphasizes an individual's right not to be tracked within stores or after products are purchased and provides some "acceptable" uses of RFID technology for tracking products in the supply chain.

The three part framework includes:

- Technology Assessment—Privacy advocates suggest that RFID technology undergo a multi-disciplinary formal technology assessment process that includes participation by all stakeholders, including consumers, which is conducted by a neutral third party.
- Principles of Fair Information Practice—Privacy advocates recommend that RFID technology be guided by strong principles of fair information practices and that minimum guidelines should be adhered to.
- Openness or Transparency—RFID users should make public their policies and practices involving the use and maintenance of RFID systems, and there should be no secret databases. Individuals have a right to know when products in the retail environment contain RFID tags and readers. They also have the right to know the technical specifications of those devices. Labeling must be clearly displayed and easily understood. Any tag reading that occurs in the retail environment must be transparent to all parties.
- Purpose Specification—RFID users should give notice of the purpose for which tags and readers are used.
- Collection Limitation—The collection of information should be limited to that which is necessary for the purpose at hand.
- Accountability—RFID users are responsible for implementation of the technology and the associated data. RFID users should be legally responsible

for complying with principles and an accountability mechanism must be established. There should be entities in both industry and government to whom individuals can complain when provisions have been violated.

- RFID Prohibited Practices

  Merchants should be prohibited from forcing or coercing customers into accepting live or dormant RFID tags in the products they buy.

  There should be no prohibition on individuals to detect RFID tags and readers and disable tags on items in their possession.

  RFID tags must not be used to track individuals absent informed and written consent of the data subject.

- Acceptable Uses of RFID Technology

  Tracking Pharmaceuticals—From point of manufacture to the point of dispensing to deter counterfeiting, and ensure proper handling and dispensing.

  Tracking Manufactured Goods—From the point of manufacture to the location where they will be shelved for sale to deter loss or theft as they move through the supply chain. Tags should be confined to the outside of products packaging and be permanently destroyed before consumers interact with the product as they leave the store.

  Detection of Items Containing Toxic Substances—When they are delivered to the landfill.

In sum, privacy advocates are requesting a review by manufacturers and retailers on item-level RFID tagging.

8.2.3 The RFID Industry Responds to Privacy Concerns

With the drive to place RFID item-level tags on consumer products, the RFID industry was focused on testing the technology and had limited awareness of the public policy issues implications associated with RFID. In response to the privacy concerns created by the initial consumer product testing, EPCglobal, the not-for-profit industry organization that is building the global EPC RFID network, formed a public policy steering committee to examine how to balance consumer privacy concerns with the industry's progress and practices.

EPCglobal adopted policy guidelines aimed at protecting consumer privacy. The guidelines "are intended to complement compliance with the substantive and comprehensive body of national and international legislation and regulation that deals with consumer protection, consumer privacy and related issues. They are based, and will continue to be based, on industry responsibility, providing accurate information to consumers and ensuring consumer choice."[55] The guidelines are as follows:

---

[55] *Guidelines on EPC for Consumer Products* (www.epcglobalinc.org public policy/public policy guidelines), EPCglobal

- Consumer Notice—Consumers will be given a clear notice of the presence of EPC on products or their packaging. This notice will be given through the use of an EPC logo or identifier on the products or packaging.
- Consumer Choice—Consumers will be informed of the choice that they have to discard, disable, or remove the EPC tags from the products they acquire. It is anticipated that, for most products, the EPC tags would be part of disposable packaging or would be otherwise discardable. EPCglobal, among other supporters of this technology, is committed to finding additional cost-effective and reliable alternatives to further enable consumer choice.
- Consumer Education—Consumers will have the opportunity to obtain accurate information about EPC and its applications, as well as information about advances in the technology. Companies using EPC tags at the consumer level will cooperate in appropriate ways to familiarize consumers with the EPC logo and to help consumers understand the technology and its benefits. EPCglobal would also act as a forum for both companies and consumers to learn of and address any uses of EPC technology in a manner inconsistent with these guidelines.
- Record Use, Retention, and Security—As with conventional bar code technology, companies will use, maintain, and protect records generated through EPC in compliance with all applicable laws. Companies will publish, on their websites or otherwise, information on their policies regarding the retention, use, and protection of any consumer specific data generated through their operations, either generally or specifically with respect to EPC use.

The purpose of the guidelines is to provide a basis for the use of EPC tags on consumer items. It is recognized by EPCglobal that for RFID technology to gain broad acceptance, consumers must have confidence in its value, benefits, and integrity of use, and modifications to the guidelines will presumable evolve as the technology is developed and implemented.

8.2.4   A Note on the "Kill Switch" Alternative

The most straightforward approach for the protection of consumer privacy is to "kill" RFID tags at the point of sale before they are placed in the hands of consumers. A killed tag is truly dead and can never be re-activated. For example, a checkout clerk in a supermarket would "kill" the tags on purchased goods and no goods would contain active RFID tags after purchase.

Reacting to early privacy concerns generated by consumer testing, EPCglobal incorporated a kill switch feature into RFID specifications. This allows the consumer to deactivate the RFID tag upon leaving a store.

8.2.5 Legislation and Regulation

Several states have had legislation introduced to set limits on the use of RFID technology. In addition, a congressional subcommittee has also held hearings on RFID technology and the Federal Trade Commission has convened a workshop on the topic to determine if federal regulation is necessary. Most of the proposed state legislation addresses the issue of ensuring notice to the consumer that RFID tags are on the product for sale and establishing policies and guidelines relating to RFID tags.

Privacy advocates, on the other hand, are pushing for legislation on RFID technology. They have drafted sample federal legislation that would outlaw some RFID devices and limit the collection of personally identifiable information.

Ideally, self-regulation by the RFID technology industry would be preferable to government legislation and regulation. However, the RFID industry, through EPCglobal, has moved quickly to improve its response to consumer privacy concerns. Although the RFID industry is still in an early stage of experimentation, it needs to continue to respond to consumer privacy concerns by clearly defining the scope and the limitations of the information it gathers and the dissemination practices it intends to follow as it develops and deploys RFID technology at the consumer product level.

8.3 THE COSTS OF DEVELOPING AND DEPLOYING RFID TECHNOLOGY

One of the major challenges inhibiting widespread use of RFID technology is the cost of RFID tags. Today, RFID tags cost between $0.30 and $0.60. For luxury products, $0.50 per unit RFID tags can easily be absorbed. However, for use in lower cost consumer products, a $0.50 tag on a tube of toothpaste would be prohibitive. A $0.05 RFID tag appears to be the benchmark tag price that industry informally agrees will lead to ubiquitous use of RFID technology. Although the price of passive UHF RFID tags will drop dramatically during the next four years, the degree of adoption of RFID technology will depend on how low the price drops. Consequently, the cost of RFID tags will be a major inhibitor to increased usage of RFID. For this reason, it appears that the retail industry's next tier of RFID technology development and application will be tagging high-end items with RFID tags that cost in the $0.25 or less range.

The costs associated with RFID systems are not limited to just the tags, however. The hardware and software needed to build an RFID system can still be very expensive too, not only to buy but to make as well. In addition, because RFID is still a developing technology, some suppliers and end-users might perceive a high degree of risk associated with entering the RFID arena and therefore be reluctant to make a large capital investment in the technology just yet.

In the near-term vendors interested in selling their products to early RFID adopters (Wal-Mart, Target and DoD) will have to embrace RFID technology to remain in the marketplace, regardless of the traditional ROI calculations.

The time and manpower that end-users will have to invest in training employees on new RFID systems is one more cost that could inhibit the adoption of RFID technology.

8.4 THE GROWTH OF GLOBAL STANDARDS AND REGULATIONS

All retailers are able to read bar codes because there are global standards, including a numbering system. However, with RFID, there are many different types of tags and different methods of communication.

The lack of uniform regulations has also been an obstacle to the adoption of RFID. Because national governments have been responsible for RFID spectrum allocation, there is international variation in the frequencies and power levels available to RFID systems. As a result, systems produced in one country (China) may not necessarily work in another. The main differences among countries are in the UHF band, which is currently the band of greatest interest to the RFID industry and the band at which most technological innovation is taking place. If world regulatory bodies are not able to agree on a more uniform set of RFID regulations, interoperability between systems around the world will remain low, that could inhibit the adoption of RFID, particularly in global supply chain applications.

There are a few reasons why globally accepted (China, India, etc.) RFID standards and regulations have not yet been fully adopted. Fighting amongst various standards and regulatory groups is one of them. Vendors have been reluctant to give up the royalties they collect on proprietary systems and move to standardized technology as well. This lack of standards and regulations, and a lack of competition amongst vendors, has meant a slower start than need be for RFID technology.

The good news is that the EPC global RFID technology standards have been recognized by the International Standards Organization (ISO) in 2006. The EPCglobal UHF Generation 2 protocol for radio frequency identification (RFID) has been endorsed by the International Standards Organisation (ISO), paving the way for its use throughout the global supply chain.

EPC is an international trade standard designed to drive RFID use forward in the UHF (ultra high frequency) range. The standard was developed so that manufacturers are using compatible devices and RFID technologies.

The royalty-free standards developed by EPCglobal are the foundations in the continuing construction of a global supply chain information network that combines RFID technology, existing communications network infrastructure and a system called Electronic Product Code (EPC), a number for uniquely identifying an item.

A unified data system would allow changes in information about product sizes, weight, name, price, classification, transport requirements and volumes to be immediately transmitted along the supply chain. For example it would allow shippers to immediately know if the amount of product stacked on a pallet had changed, or give a retailer time to adjust display space.

The system is being built to help companies save money throughout the supply chain by using the Global Data Synchroization Network (GDSN). Nestle, Coca-Cola, PepsiCo, Hormel Foods, Kraft, Unilever, Wegmans Food Markets, and Sara Lee are among the food companies that have signed up to implement the system.

In a boost for the standard, EPCglobal has announced that the ISO has incorporated its Generation 2 RFID air interface protocol into its ISO/IEC 18000-6 Amendment 1 as Type C on UHF RFID.

About a dozen RFID readers, tags and integrated circuits have been certified as Gen 2 compliant by EPCglobal and are commercially available.

The standard was initially developed by more than 60 technology companies and describes the core capabilities required to meet the performance needs set by the end user community.

- Broaden the market for RFID
- Give rise to products and applications interoperability
- Reduce development and manufacturing costs
- Promote technology acceptance and technology advancement

Not until RFID is fully standardized will the industry be able to realize all of these goals. The standards that the International Standardization Organization (ISO) and EPCglobal have developed should fill the need. Some anticipate that the EPC Class I Generation 2 standard will resolve the standards problem.

8.5 TECHNOLOGICAL IMMATURITY AND INTEGRATION WITH LEGACY SYSTEMS

Non-mandated RFID use by global business enterprise interest in RFID technology is still relatively new. In addition, software to integrate RFID technology with ongoing business applications (middleware) is also very immature. Currently, software companies (Microsoft, Oracle, IBM, etc) are making considerable investments to integrate RFID technology with business applications but this will take several years to mature. Until the technology has matured, widespread adoption is still unlikely.

Even if all RFID tag and reader issues are worked out, this won't produce the real-time flow of information technology systematic data that companies need to gain the full benefits of RFID technology. RFID is going to change business processes in such a way that users will have to either install new applications or endure a complex rewrite of existing programs. This will take a great deal of time and presents another obstacle to the widespread use of RFID.

8.6 LACK OF ROBUSTNESS

8.6.1 RFID Accuracy

RFID accuracy refers to the success rate at which a reader can identify a single tag that enters its read zone. Accuracy is affected by many things and identification can be marred by a number of physical constraints. They include:

- Reader interference—Reader collision can have a deleterious affect on accuracy. Signals from different readers can overlap and interfere with one another.
- Environment—A number of different environmental factors can affect accuracy. Objects in the environment of the readers and the tags affect both high and low frequencies, particularly metal objects. Higher frequencies are easily absorbed by water.
- Tag Orientation—Tag orientation can also lower RFID accuracy. The presumably random arrangement of tags in a read zone could render some tags invisible to the reader.
- Distance and Power—The variability in distances between tags poses problems for systems designers and can reduce accuracy. Wide variations in power when signals propagate through various materials can also reduce accuracy.

All of the above problems are inherent in today's RFID system. In today's RFID systems, however, which are not very mature and therefore not very robust, they can severely inhibit accuracy. It can be assumed that someday RFID systems will be able to effectively deal with all of these problems and operate at a high degree of accuracy, but until then RFID accuracy will pose a barrier to the widespread adoption of RFID.

8.6.2 Scalability

RFID scalability is the rate at which a single RFID reader is able to successfully identify a large number of tags simultaneously. Whereas accuracy is adversely affected primarily by physical constraints, scalability is affected by limitations on the computing power of an RFID interrogator and the network it is connected to. In order to produce scalability tag collision must be effectively handled. The anti-collision measures used in some of today's RFID systems are not very robust. It will take some time until they are and RFID systems are able to obtain a high degree of scalability. Until then, the lack of robustness will inhibit adoption of the technology.

8.7   LACK OF KNOWLEDGE AND EXPERIENCE, END-USER CONFUSION, AND SKEPTICISM

In comparison to the use of barcodes, RFID technology is still a complex technology in which little experience has been gained. Knowledge of the technology is relatively low in most organizations and installation of RFID technology currently lies with small companies that are involved in the initial projects and installations. Before there is widespread development of RFID technology it will require the participation, support, knowledge, and expertise of larger technology development companies.

There is also a great deal of confusion surrounding RFID technology. This is due in part to the marketing strategies of some companies that develop RFID systems. In their effort to generate interest in RFID, a great deal of hype has been created. As a result, there are some misperceptions about just what the technology is able to deliver now and what it will be able to deliver in the future? Furthermore, some companies have marketed RFID as a bar code replacement technology, which many will contend it is not. While RFID can compete with bar codes in many ways, and the benefits of RFID often surpass those offered by bar codes and justify the costs, RFID systems are vastly more expensive, more complicated and less robust than bar code systems. When some customers finally realize this, and the costs involved with RFID, for example, that while bar codes cost a penny a piece an RFID tag often costs 50 times that, they have become disappointed and skeptical of the technology.

8.8   ETHICAL ISSUES

Ethical questions also pose an obstacle for some RFID applications. For example, in 2004, FDA approved the human use of the implantable VeriChip as a medical device for patient identification and health information and that it could be used "only to store a unique electronic identification code that is used to access a patient's identification and corresponding health information stored in a database." This FDA ruling gave the green light to Applied Digital Solutions (ADS) to commercially market its VeriChip to the healthcare industry and the public as a personal identification and medical record storage device.

The FDA approval of the human implant VeriChip raised many ethical, access, and data security concerns. Groups opposed to "chip" implants cover the full range of the national political, religious, and social spectrum. Some groups object to the procedure on purely religious grounds while others see it as an assault on individual liberty and personal privacy.

Concerns have also been expressed about the access and security of the personal identification and medical record data maintained in the ADS VeriChip healthcare database. Ethical, access, and data security and safety concerns include:

- Does the person/reader have the proper authorization to access the ADS-maintained medical record. Has proper authorization been granted to the person/reader. What if the individual is unconscious in an emergency situation and is unable to authorize access?
- Will use of RFID tags ("chipping") become a prerequisite for membership in a HMO or other health insurance plan.
- Is the information in the ADS-maintained medical record database current and accurate. How is information in the ADS database updated? Must healthcare providers subscribe to the VeriChip program to update information.
- FDA also expressed concerns about the safety aspects of an MRI scan on the implanted chip (metal heats up when subjected to MRI). Is this an important healthcare safety hazard issue?
- Finally, the VeriChip personal identification code is intended for medical use but could possibly be co-opted by unscrupulous others as a new and unique method of identification theft.

Clearly, the answers to these questions are not currently available and the debate will continue to revolve over whether the benefits of RFID implants outweigh the concerns and whether sufficient and satisfactory actions will be taken by various levels of government, the industry and individual companies to allay the ethical, data access, security, and safety concerns of human RFID implants.

8.9   DATA MANAGEMENT

In the effort to address these many issues, adopters of RFID technology are overlooking a seemingly mundane but important aspect of RFID deployment: making sure back-end databases and business applications can handle the massive amounts of new data that RFID systems will produce. In the rush to implement RFID, users are overlooking the implications to their IT systems. Too much focus is placed at present on the price of the tags and abilities of readers and not enough on the data and how it's going to be used. If IT infrastructures are not updated to handle the new load they will suffer and shaky infrastructures could collapse.

The following is a list of challenges that adopters face when managing RFID data[56]:

- Large Volumes of Data—RFID systems will have an unprecedented ability to produce great volumes of raw data in relatively short periods of time. Adopters of RFID technology must ensure their IT systems are dimensioned accordingly.

---

[56] *Microsoft and RFID: Microsoft White Paper*, Microsoft, September 2004.

- Data Integration Across Multiple Facilities—Enterprises with geographically distributed facilities networked to a central IT facility will be faced with the problem of managing raw RFID data while at the same time aggregating it into the central IT facility. Having large quantities of data flowing across network interconnects could place a burden on those enterprises' IT infrastructures.

- Data Ownership and Partner Data Integration—In retail supply chains or other applications in which data would need to be shared between different companies, questions might arise pertaining to the ownership of data. This could hinder integration of RFID systems between the companies.

- Product Information Maintenance—In some applications, retail supply chains for instance, central IT databases might continually need to be accessed to retrieve product information. In large scale implementations, when a high volume of tags are processed, this could put a burden on IT infrastructures.

CHAPTER 9

THE FUTURE PREDICTIONS FOR RFID

Interest in RFID technology is growing rapidly. The Wal-Mart and DoD initiatives, the quickly falling costs of implementing the technology, the emergence of an RFID standard, and a potentially high return on investment are all contributing factors to this phenomenon.

A number of companies across a wide rage of industries, and government organizations as well, are seeking to increase operational efficiency, lower operating costs, and/or increase profits through the use of RFID technology. More and more RFID pilot trials and mandates are being announced every month as a result.

Over the next five years, the RFID industry will experience explosive growth, both in terms of dollar sales and applications available. In 10 to 15 years RFID technology will be ubiquitous.

This book discussed the technical characteristics of RFID, the history of RFID technology, several applications of the technology with a focus on the commercial supply chain, government, law enforcement, and corrections applications, pharmacy and many of the issues involved in the widespread deployment of RFID.

RFID systems are composed of three basic building blocks: tags, readers, and hosts. RFID tags come in a variety of forms and types and their high price has inhibited the widespread adoption of RFID technology. Prices are falling quickly, however, and by the end of 2007 they are expected to cost as little as

RFID-A Guide to Radio Frequency Identification, by V. Daniel Hunt, Albert Puglia, and Mike Puglia
Copyright © 2007 by Technology Research Corporation

$0.05 a piece. RFID readers are responsible for communicating with RFID tags and relaying information to and from RFID host computers. They also implement security and anti-collision measures. Finally, RFID hosts are used to network multiple RFID readers together to form coherent RFID networks. They also direct RFID data to and from enterprise IT networks.

Enterprise IT networks and middleware software development and adoption are central to business and decision-making processes. Thus, in order to make use of RFID data in these processes, RFID networks need to be integrated with enterprise IT networks. RFID middleware is used to do this. Middleware routes data between RFID and IT networks. It is ultimately responsible for the quality and usability of RFID data. It has four main functions: data collection, data routing, process management, and device management. Major software corporations such as Microsoft, Oracle, IBM etc. Are aggressively developing RFID IT middleware tools.

RFID has been called a replacement technology for bar codes. Critics call the comparison inappropriate, citing the much higher cost of RFID technology. RFID offers many capabilities that bar code systems cannot, however, such as the ability to both read and write to tags, the ability to operate without a direct line of sight between tag and reader, and the ability to communicate with hundreds of tags simultaneously rather than one at a time. These capabilities can produce cost-saving benefits that will offset the high price of implementing RFID in many cases.

In the late 1980s, automatic toll collection systems appeared on the market, followed by point-of-sale systems such as ExxonMobil's Speedpass in the 1990s. Towards the end of the 1990s, advances in materials science research put cheap RFID tags on the horizon and many venture projects were started with the aim of applying RFID to supply chain and asset management problems. The Auto-ID Center, now part of EPCglobal and makers of the EPC standard, was started during this period as well. Finally, in 2003, the Wal-Mart and DoD RFID mandates gave momentum to the early adoption of RFID.

In the near-term commercial applications of RFID technology can be broken down into four categories: retail and consumer packaging, transportation and distribution, industrial and manufacturing, and security and access control. While RFID penetration is increasing across all application segments, supply chain and asset management applications are projected to lead the industry's growth for the foreseeable future. RFID offers capabilities that no existing technology can, such as complete supply chain visibility and item-level tracking of merchandise. These capabilities can provide higher operating efficiencies and lower operating costs to the organizations that use RFID.

Governments are also seeking and finding ways to leverage RFID technology to improve services and efficiency and to lower operating costs. The Department of Defense is currently the leader in government use, though many other federal agencies have begun their own projects, including the Food and Drug Administration, General Service Administration, and the

Department of Homeland Security. State and local governments are seeking ways to use RFID technology in airport, transportation, and corrections applications.

The Department of Homeland Security will use RFID technology to support the identity management and location determination systems that are fundamental to controlling the U.S. border and protecting transportation systems. RFID will be combined with other technologies, such as GPS and biometric systems, to create "smart borders" and to secure international shipping containers. The U.S.-VISIT program is the first such application.

Law enforcement applications of RFID have been slow to develop in the United States due to privacy concerns and a lack of awareness amongst law enforcement organizations. Several law enforcement applications of RFID have been identified however. They are designed to improve police efficiency and ensure officer safety and to employ RFID as a crime fighting , forensic and investigative tool. Forensic science (CSI), evidence tracking, and property control systems and police tracking devices are a few examples.

RFID systems will be used in corrections to allow continuous inmate tracking to prevent escape, reduce violence, and continuously monitor and record the location of inmates and guards within the prison. Several pilot projects have been completed successfully and many corrections systems are beginning to take notice.

While the future of the RFID industry looks promising, there are still many issues to be overcome before the technology will be widely adopted. RFID technology is not widely understood. This is both due to the fact that it is so new, as well as the marketing practices of the RFID industry. If RFID is perceived as a high-risk investment because of this uncertainty, it will inevitably delay the widespread adoption of the technology and slow the growth of the industry.

A lack of globally accepted regulations and standards is also inhibiting the deployment of RFID, particularly in global supply chains. The FCC and the regulatory agencies of Europe, China, and Japan have not yet come to agreement on several issues, particularly the UHF band and effective radiated power levels for RFID readers.

The high cost of RFID technology has been a barrier to the industry's growth as well. Cheaper RFID tags will go a long way towards lowering this barrier. However the DoD and Wal-Mart RFID adoption mandates will drive the cost down. The predominance of proprietary systems is another cause for this barrier, and technology standards will play a significant role in bringing prices down.

Finally, the widespread adoption of RFID technology has alarmed privacy advocates. The tracking capabilities that RFID technology can provide, both for merchandise and people, and the potential abuse of this power, is a source of concern.

The economic benefits of deploying RFID technology can be summarized as follows:

- Cost Reduction—The cost reduction value case is the goal of many consumer packaged goods companies, retailers, and the U.S. Department of Defense. These enterprises expect to reduce inventory and inventory management expenses by billions of dollars over the next several years through RFID deployment.
- Increased Revenue—Both large and small retailers and manufacturers are developing RFID deployments to drive sales. RFID can increase revenue by reducing out of stock items and materials, reducing item shrinkage, and improving inventory operations.
- Counterfeit Product Shielding—Manufacturers lose sales and profits from the flow of counterfeit products, such as high dollar valve drugs. Many of these products also present safety and security hazards for customers. RFID can effectively eliminate this problem.
- Shrinkage, Theft, and Diversion—High-value consumer and industrial products face the large risk of theft and diversion. RFID has been shown to reduce theft and diversion from the store shelf and the supply chain, or from the factory floor to the storefront.

RFID is here to stay. In the coming years, RFID technology will slowly penetrate many aspects of our lives, just as television, personal computers, and mobile phones already have. Those companies and government organizations that decide to research and invest in the technology now will not only become the early winners but also derive a benefit from their early knowledge when extending the technology to new applications in the future.

APPENDIX A

WAL-MART RFID INITIATIVE

Wal-Mart launched its RFID initiative on June 11, 2003, when it issued its first RFID mandate for suppliers. In that mandate, Wal-Mart formally announced that it would require its top 100 suppliers to begin tagging pallets of merchandise by January 2005, and all suppliers were to begin tagging pallets by January 2006. A few months after issuing the mandate, in September of 2003, Wal-Mart then opened its own RFID lab, which has been tasked with researching ways to apply RFID technology to Wal-Mart operations and with formulating Wal-Mart's RFID policy.

Pilot testing began in October 2003. A specialty distribution center and two suppliers were used in the initial run. In late 2003, Wal-Mart also began an on-going effort to communicate RFID policy to both suppliers and technology vendors when it held an "RFID Symposium" with its top 100 vendors and by participating in an RFID trade-show.

Wal-Mart has decided not to make its full RFID policy available to the public. Instead, Wal-Mart has chosen to distribute the full text of the policy to its suppliers only, via its "Retail Link," which is a Wal-Mart-designed, web-based IT application used to communicate with suppliers. The bulk of policy information made available to the public comes by way of a few Wal-Mart press releases, however, there is a limited amount of information available from connected third parties also.

This is what is known about the Wal-Mart RFID initiative and Wal-Mart's RFID policy, through Wal-Mart press releases and connected third parties:

RFID-A Guide to Radio Frequency Identification, by V. Daniel Hunt, Albert Puglia, and Mike Puglia
Copyright © 2007 by Technology Research Corporation

- The Wal-Mart RFID mandate specifies the use of EPC Class 0 and Class 1, 96-bit Gen 1 tags. Sixty-four-bit tags are not being supported. Wal-Mart has stated that they are driving towards the implementation of the Gen 2 EPC standard, that is now available.
- Wal-Mart is requiring one antenna on each side of dock door/portals; one antenna above dock doors; and one antenna on each side or underneath conveyors moving up to 600 ft/min for case tagging. Also, cases have to be read with 100% accuracy at 540 ft/min.[57]
- In April 2004, RFID went live in Wal-Mart stores for the first time. This pilot project was located in North Texas and included 21 products from 8 suppliers and 7 local stores. The suppliers were: The Gillette Co., HP, Johnson & Johnson, Kimberly-Clark, Kraft Foods, Nestle Purina PetCare Co., The Proctor & Gamble Company, and Unilever.
- In August 2004, Wal-Mart announced that significant RFID expansion would occur over the following 16 months.
- During 2005 continued to expand its use of RFID to improve their supply chain.
- In 2006 Wal-Mart expanded its RFID initiative to an additional 300 prime vendors.

That said, there is more about Wal-Mart's RFID that can be gleaned from other sources.[58]

While Wal-Mart has championed the RFID cause, it has not stated that it will abandon bar code technology. Bar codes are far too pervasive and important to supply chain management at present to contemplate doing so. Wal-Mart has, however, been quick to point out the advantages that RFID provides over existing bar code systems, perhaps indicating that their long-term vision is to fully replace bar codes with RFID.

In describing the merits of RFID technology, Wal-Mart has pointed out the following advantages that it provides over bar code systems:

- RFID does not require line of sight for scanning
- RFID can perform in harsh, rugged environments, where bar codes cannot
- RFID labels do not present space issues, as they can be hidden inside packaging, whereas bar codes can consume a great deal of space on small items
- Bar codes have a limited capacity for storing information, whereas RFID tags have, for all intents and purposes, unlimited capacity
- RFID has read/write capability, whereas bar codes do not

[57] http://www.prweb.com/releases/2003/11/prweb88946.htm.
[58] Wal-Mart RFID Presentation, Simon Langford, ISD RFID Strategy, 2003.

- Because the information on bar codes can not be altered, that information is doomed to become "stagnant" and out-dated over time, whereas RFID tags can be updated as necessary. Also, because bar codes cannot be uniquely identified, when a bar code reader scans three identical bar code tags, it has no way of determining whether the same item has been scanned three times or whether three separate but identical items have been scanned.
- Multiple RFID tags can be read simultaneously whereas bar codes cannot.
- With bar codes, which often have to be manually scanned or positioned in front of a bar code reader, there is greater opportunity for human error, for instance, missed scans or scanning the same item multiple times.

Having stated these advantages to RFID, it can be assumed that these are the ways in which Wal-Mart believes RFID technology will most significantly impact their business.

From an operational standpoint, in deploying RFID technology, Wal-Mart's primary aim is to enable greater supply chain visibility. Wal-Mart's RFID vision falls into line with EPCglobal's vision; that is, RFID will impact the entire supply chain, from manufacturer/supplier to the distribution center to the retail level. Wal-Mart, like EPCglobal, has identified a four-tiered distribution cycle:

- In the first tier, RFID tags are placed on items, cases, and pallets at the manufacturer/supplier's facility. When merchandise is shipped from the factory, it begins being tracked through RFID.
- In the second tier, tagged merchandise is received at distribution centers and continues to be tracked by RFID systems installed at those facilities. Merchandise is tracked when it enters warehouses, throughout the course of its storage and even when being moved within warehouses, and again upon exiting the facility, when it is shipped to retail locations.
- In the third tier, tagged merchandise arrives at retail locations. RFID systems installed in the store backroom, storage racks, and sales floor shelves track the merchandise throughout its retail life.
- In the fourth tier, which Wal-Mart recognizes will only exist at some time in the distant future, RFID will enable streamlined customer checkout. When merchandise is "bought" by a customer, upon exiting the store sales receipts can be written up and payment made without the traditional checkout and "buying" process having to take place.

While supply chains overall stand to benefit from the application of RFID technology, the different members of supply chains—suppliers/manufacturers, distribution centers and stores—stand to benefit in different ways. Wal-Mart has identified some of the ways in which RFID will potentially do this:

- Suppliers/Manufacturers—RFID systems and the information they can collect about product demand will enable suppliers to plan production more efficiently, according to Wal-Mart. Reduced inventory will be the result, as well as improved inventory control. Furthermore, RFID will enable "smart" recalls, by targeting defective lots more accurately. Faster shipping and receiving—which is a benefit enjoyed throughout the supply chain—will be enabled through RFID also.
- Distribution Centers—Retail distribution centers will benefit primarily through automated inventory counts, according to Wal-Mart. Improved quality inspection at distribution centers will be enabled through the use of RFID technology, ostensibly due to the amount of labor that will be freed from having to conduct time-consuming inventories and redirected towards the inspection process. And again, retail distribution centers, like the rest of the supply chain, will benefit from faster shipping and receiving.
- Stores—RFID will benefit stores primarily through reducing stockouts. According to Wal-Mart and an Emory OOS (Out-of-Stock) study conducted in 2002, a typical retailer loses about 4% of sales due to out-of-stock situations. Improved customer in-stock, enabled through RFID, will lower these costs. Theft prevention, lower shrink, and automated checkout are several other potential store benefits that Wal-Mart has cited. And finally, as with the retail distribution centers, RFID will reduce inventory and enable real-time inventory at the store level.

Wal-Mart's stated short-term focus is on solutions that will deliver an immediate return on investment to both Wal-Mart and its suppliers. As a result, RFID tagging will be done at the case-and-pallet level initially, with item-level tagging to be delayed for some time.

Wal-Mart has broken down its domestic distribution center operations into three different levels. They are:

- RDC (Regional Distribution Centers)
- GDC (Grocery Distribution Centers)
- Sam's Club (Dry Cross-Dock Distribution Centers)

The RFID mandate applies to all three levels. Furthermore, Wal-Mart has classified its retail operations into three categories, all of which are part of the RFID mandate:

- Wal-Mart Discount and Super-Centers
- Sam's Club
- Neighborhood Markets

While Wal-Mart has endorsed the EPCglobal standard in its mandate, it has specifically not endorsed any particular technology provider. Their goal for endorsing the EPC standard was to promote competition and drive costs down, and endorsing any particular technology provider would run contrary to this. EPCglobal has launched a provider "certification" initiative in which it will certify that technology providers are in compliance with EPCglobal standards. Wal-Mart is supporting this initiative and is apparently willing to do business with any provider that is certified through EPCglobal.

Moving forward, Wal-Mart has identified the following milestones in its RFID timeline:

- **2004**

    Test pallet-level implementation strategy
- **2005**

    Top 100 suppliers begin tagging pallets in January

    RFID live in six distribution centers and 250 stores by June

    RFID live in 13 distribution centers and 600 stores by October

    Rollout of pallet-tagging to include next 200 suppliers by end of year
- **2006**

    Item-level tagging begins, with tags costing $0.25 or less, of merchandise including: tires, electronics, pharmaceuticals, high theft items, high ticket items and case items

    Expanded item-level tagging, with tags costing $0.05 or less, of increasingly cheaper items

    International expansion

A press statement issued by Wal-Mart's CIO, Rollin Ford, said basically that the Wal-Mart RFID program is still on track, with the next wave of 300 suppliers expected to be using RFID for cases and pallets in January 2007. Another 500 Wal-Mart RFID capable products will be added by the end of the year.

Wal-Mart is simply reaffirming its RFID/EPC (electronic product code) program, and touting the benefits. The Wal-Mart schedule will ultimately impact the various "tipping points" for major suppliers.

The Wal-Mart press release stated that by the end of the year, more than 1000 of its traditional and Sam's club stores will be RFID-ready. That would represent nearly 25% of the companies U.S. stores.

In addition, the company said that from this point forward, the company will read "Gen 2" EPC tags.

"Recent internal analysis of Wal-Marts ongoing efforts, along with the launch of Generation 2 tags, reinforces the value of this technology for Wal-Mart and ultimately our customers," said Rollin Ford, executive vice president and chief information officer for Wal-Mart, and former head of supply chain before taking on his new role. "We're aggressively moving forward with the Wal-Mart RFID-enabled facilities."

The new wave of 300 suppliers is expected to start testing tag shipments in the coming months, and be live in January 2007.

Ford continues to tout the benefits not only for Wal-Mart but suppliers and plans to work with suppliers to help them see the vast potential of RFID." Ford is already fully convinced of its value and is ready to step up the pace of RFID use at Wal-Mart.

In conclusion, Wal-Mart will continue to help drive the adoption of RFID supply chain applications for the foreseeable future. Many of its policies will ultimately be determined in concert with the EPCglobal RFID inititatives.

Note: The above description of the Wal-Mart RFID initiative is the author's view of their efforts based on public domain, and internet non-copyrighted text. Wal-Mart has not edited or endorsed this description of their RFID initiative.

DEPARTMENT OF DEFENSE RFID POLICY OVERVIEW

The Department of Defense has described its "Radio Frequency Identification (RFID) Policy" in portions of the following memorandium by the Under Secretary of Defense, initially issued on July 30, 2004:

As the Defense Logistics Executive (DLE), this memorandum issues the policy for implementing Radio Frequency Identification (RFID) across the Department of Defense (DoD). This policy finalizes the business rules for the use of high data capacity active RFID and finalizes the business rules for the implementation of passive RFID and the use of Electronic Product Code™ (EPC) interoperable tags and equipment (EPC Technology) within the DoD supply chain and prescribes the implementation approach for DoD suppliers/ vendors to apply passive RFID tags. This policy memorandum applies to the Office of the Secretary of Defense (OSD); the Military Departments, the Joint Chiefs of Staff and the Joint Staff; the Combatant Commands; the Inspector General of the Department of Defense; the Defense Agencies, and the DoD Field Activities (hereafter referred to collectively as the "DoD Components"). An internal implementation strategy for DoD Components to read and apply passive RFID tags will be issued in a separate Defense Logistics Executive (DLE) decision memorandum. This policy supersedes two previous issuances of policy dated October 2, 2003, and February 20, 2004.

DoD Components will immediately resource and implement the use of high data capacity active RFID in the DoD operational environment. DoD outlines

RFID-A Guide to Radio Frequency Identification, by V. Daniel Hunt, Albert Puglia, and Mike Puglia
Copyright © 2007 by Technology Research Corporation

the detailed guidance on active tagging. DoD Components must ensure that all consolidated shipments moving to, from, or between overseas locations are tagged, including retrograde, and must expand the active RFID infrastructure to provide global intransit visibility. In order to take advantage of global RFID infrastructure not within DoD's control, the DoD Logistics Automatic Identification Technology Office will assess the ability to leverage any compatible active RFID commercial infrastructure that commercial entities may establish. This should not be viewed as direction to commercial carriers and port operators to establish an active RFID infrastructure.

This appendix contains the detailed guidance on implementation of passive RFID capability within the DoD supply chain as well as the data constructs for the tags. DoD will use and require its suppliers to use EPC Class O and Class 1 tags, readers and complementary devices. DoD will migrate to the next generation tag (UHF Gen 2) and supporting technology. When the specification for UHF Gen 2 was finalized, the Department announced a transition plan to this technology, but we expect use of EPC Class O and Class 1 technology through 2007.

Radio Frequency Identification is a mandatory DoD requirement on solicitations issued on or after October 1, 2004, for delivery of materiel on or after January 1, 2005, in accordance with the DoD supplier implementation plan. Contracts with DoD shall require that passive RFID tags be applied to the case, pallet and item packaging for unique identification (UID) items. The Defense Logistics Board (DLB) will review the internal implementation plan, benefits, compliance requirements, and requisite budget requirements annually based on an assessment of the implementation to date. This review will include an updated analysis of implementation success as well as provide guidance for expansion of RFID capabilities into additional applications and supply chain functional processes. A DLE decision memorandum will provide funding guidance for DoD Component implementation.

In order for the DoD Components to meet the requirements of this policy, the DoD has developed a Department-wide RFID Concept of Operations (CONOPS) to outline the transformational role of RFID technology in DoD logistics and to articulate the specific uses of both active and passive RFID throughout the DoD supply chain. Components will prepare a supporting RFID implementation plan that encompasses both active and passive RFID technology in a cohesive environment to support the DoD vision.

To support the purchase of passive RFID technology and leverage the purchasing power of the Department, the Army's Program Executive Office Enterprise Information Systems (PEG EIS) continues development of a multi-vendor contract mechanism to procure EPC technology. This contract will include competitive vendors providing RFID equipment/infrastructure in accordance with current published EPC specifications (Class O and Class 1) and specifications for UHF Gen 2.

To institutionalize RFID as a standard way of doing business, this policy will be incorporated into updates of the DoD Supply Chain Materiel Management Regulation (DoD 4l40.l-R), the Defense Transportation Regulation (DoD 4500.9-R), and the Military Standard 129. Likewise, DoD Components will incorporate this policy into Service/Agency level publications as well as Component strategies to achieve compliance with the DoD Business Enterprise Architecture Logistics (BEA-LOG).

The following policy also applies to take full advantage of the inherent life cycle management efficiencies of this technology. Beginning in FY 2007 and beyond, only RFID capable AIT peripherals (e.g., optical scanners, printers used for shipping labels) will be acquired when these peripherals support RFID capable business processes. Beginning in FY 2007 and beyond, logistics automated information systems (AISs) involved in receiving, shipping, and inventory management will use RFID to perform business transactions, where appropriate, and AIS funding will hinge on compliance with this policy.

Managers of all major logistics systems modernization programs will update appropriate program documentation to include the requirement for RFID capabilities as part of system operational deployment in conformance with the business rules and initial time line set forth in this policy. Managers of major acquisition programs will update programs as required to include the requirement for RFID capabilities where applicable. The DLB will review these requirements prior to FY 2007 implementation.

DoD will continue to work together with suppliers on this critical initiative. RFID remains part of the larger suite of AIT technologies and the Department will leverage all of these technologies, where appropriate in the supply chain, to improve the ability for DoD to support the warfighter. However, an RFID-capable DoD supply chain is a critical element of Defense Transformation and will provide a key enabler for the asset visibility support down to the last tactical mile that is needed by our warfighters. Support of the DoD RFID efforts are vital to our success in meeting this requirement. For further information, please refer to the DoD websites at www.DoDrfid.org. and www.acq.osd.mil/log/RFID.

BUSINESS RULES FOR ACTIVE RFID TECHNOLOGY IN THE DOD SUPPLY CHAIN

Overview

Active Radio Frequency Identification (RFID) tags used in DoD are data rich and allow low-level RF signals to be received by the tag, and the tag can generate high-level signals back to the reader/interrogator. Active RFID tags can hold relatively large amounts of data, are continuously powered, and are normally used when a longer tag read distance is desired.

The DoD Logistics Automatic Identification Technology (LOG-AIT) Office is the DoD focal point for coordinating overarching guidance for the use of AIT within DoD. The Program Executive Office, Enterprise Information Systems (PEO EIS), Product Manager-Automatic Identification Technology (PM-AIT) Office is the DoD procurement activity for AIT equipment (to include RFID equipment and infrastructure) and maintains a standing contract for equipment integration, installation, and maintenance. The Defense Logistics Agency (DLA) is the procurement activity and single manager for active RFID tags. Users will coordinate RFID equipment/infrastructure procurement through the PM-AIT Office and tag procurement from DLA to ensure interoperability and compliance with this policy.

The following business rules are applicable to all DoD Components. They support asset visibility and improved logistic business processes throughout the DoD logistics enterprise. These rules specifically apply to DoD cargo shipped outside the continental United States (OCONUS), however, organizations are encouraged to employ the use of active RFID technology for intra-continental United States (CONUS) shipments to support normal operations or for training.

Active RFID Business Rules

Sustainment/Retrograde Cargo   All consolidated sustainment or retrograde shipments (RFID Layer 4 freight containers (e.g., 20 or 40 foot sea vans, large engine containers, and 463L air pallets) of DoD cargo being shipped OCONUS must have active, data-rich RFID tags written at the point of origin for all DoD activities (including vendors) stuffing containers or building air pallets. Content level detail will be provided in accordance with current DoD RFID tag data specifications. Containers and pallets reconfigured during transit must have the RFID tag data updated by the organization making the change to accurately reflect current contents.

Unit Movement Equipment and Cargo   All RFID Layer 4 freight containers and palletized unit move shipments being shipped OCONUS, as well as all major organizational equipment, must have active data-rich RFID tags written and applied at the point of origin for all activities (including vendors) stuffing containers or building air pallets. Content level detail will be provided in accordance with current DoD RFID tag data standards. Self-deploying aircraft and ships are excepted.

Ammunition Shipments   All RFID Layer 4 freight containers and palletized ammunition shipments being shipped OCONUS must have active data-rich RFID tags written with content level detail. Tags will be applied at the point of origin by all activities (including vendors) that stuff containers or build air pallets in accordance with current DoD RFID tag data specifications. Containers and pallets reconfigured during transit must have the RFID tag data updated to accurately reflect current contents by the organization making the change.

Prepositioned Materiel and Supplies   All RFID Layer 4 freight containers and palletized prepositioned stocks or War Reserve Materiel as well as all major organizational equipment, must have active data-rich RFID tags written with content level detail and applied at the point of origin by all activities (including vendors). Execution for current afloat assets will be completed during normal maintenance cycle, reconstitution/reset, or sooner as required.

RFID Infrastructure USTRANSCOM will ensure that designated strategic CONUS and OCONUS aerial ports and seaports (including commercial ports) supporting Operation Plans (OPLANs) and military operations have RFID equipment (interrogators, write stations, tags, brackets) with read and/or write capability to meet Combatant Commander requirements for asset visibility. Military and commercial ports will be instrumented with fixed or mobile RFID capability based on volume of activity and duration of the requirement at the port. Military Departments and Combat Support Agencies will ensure sufficient RFID infrastructure and equipment (interrogators, write stations, tags, and brackets) are appropriately positioned to support Combatant Commander requirements for asset visibility. As above, military and commercial ports will be instrumented with fixed or mobile RFID capability based on volume of activity and duration of the requirement at the port.

To ensure that users take maximum advantage of inherent efficiencies provided by this technology, RFID capability will be operational at logistic nodes and integrated into existing and future logistics automated information systems. RFID recorded events will become automatic transactions of record. Geographical Combatant Commanders may direct Service Components/ Combat Support Agencies to acquire, operate, and maintain additional theater supporting RFID infrastructure to meet changing theater operations. As a general rule, an organization responsible for port or logistics node operation is also responsible for installing, operating, and maintaining appropriate RFID capability. Additionally, when responsibility for operating a specific port or node changes (e.g., aerial port operations change from strategic to operational), the losing activity is responsible for coordinating with the gaining activity to ensure RFID capability continues without interruption.

RFID Funding The cost of implementing and operating RFID technology is considered a normal cost of transportation and logistics and as such should be funded through routine Operations and Maintenance or Working Capital Fund processes. It is the responsibility of the activity at which containers, consolidated shipments, unit move items, or air pallets are built or reconfigured to procure and operate sufficient quantities of RFID equipment to support the operations. Working Capital Fund activities providing this support will use the most current DoD guidance in determining whether operating cost authority or capital investment program authority will be used to procure the required RFID equipment. If the originating activity of the Layer 4 container/consolidated air pallet is a vendor location, it is the responsibility of the procuring Service/Agency to arrange for the vendor to apply active tags, either by obtaining sufficient RFID equipment to provide the vendor to meet the requirement, or requiring the vendor as a term of the contract to obtain necessary equipment to meet the DoD requirement. Additionally, Combatant Commanders are responsible for coordinating with their Service Components to ensure adequate enroute RFID infrastructure is acquired and operating at key logistics nodes.

RFID Tag Return   The DLA automated wholesale management system will provide tags through existing supply channels. The DoD Item Manager for the active RFID tags (NSN 6350-01-495-3040) is the Defense Supply Center Philadelphia, Inventory Control Point, Routing Identifier Code S9I. Only new Condition Code A tags will be sold to customers. All returned tags that are serviceable after refurbishment will be received into wholesale inventory as Condition Code B and will be available as free issue from the DLA Defense Distribution Center (DDC) when they are placed on a pallet or container by DDC. This will spread the savings across the DoD Community of active tag users. When DDC requisitions tags, Condition Code B tags will be issued first. If there are no Condition Code B tags available for issue to the DDC, the DDC will pay the standard price for Condition Code A tags. Activities are encouraged to use the Defense Logistics Management Supplement Materiel Returns Program (MRP) to return tags no longer required and receive reimbursement for packaging, crating, handling, and transportation (PCH&T) costs. Excess tags sent back without MRP transactions will not result in PCH&T reimbursement to the customer. The PCH&T reimbursement incentive for tags received with MRP transactions will result in reduced costs and savings to DoD from reusing the Condition Code B tags. The Military Services, other requisitioners, and users may opt to establish their own retail operation for used tags and incur the cost of refurbishment themselves.

RFID Tag Formats   The DoD LOG-AIT Office is responsible for coordinating, establishing, and maintaining RFID tag formats at the data element level. RFID tagging procedures require active data-rich RFID tags be written with content level detail in accordance with approved formats RF Tag Data Format Specification, Version 2.0, the current version. RFID tag data files will be forwarded to the regional in-transit visibility (ITV) server(s) in accordance with established DoD data timeliness guidelines published in the current versions of the DoD 4500.9-R, Defense Transportation Regulation and Joint Publication 4-01.4, Joint Tactics, Techniques, and Procedures for Joint Theater Distribution. RF Tag data is further transmitted to the Global Transportation Network (GTN) and other global asset visibility systems as appropriate. This tag data flow will be analyzed in the future as part of the DPO architecture. RF tag formats will be identified in the current version of DoD 4500.9-R, Defense Transportation Regulation, and the format requirements will be published in MIL STD 129, DoD Standard Practice for Military

Marking for Shipment and Storage. It is the intent of the Department to incorporate all RFID tag formats and usage standards into a DoD RFID manual.

RFID ITV Server Management   The PM-AIT Office will manage the RFID ITV servers. All DoD Component operated RFID interrogators will forward their data to the ITV servers maintained by PM-AIT. This will enable the PM-AIT Office to program for funding and provide a centralized management structure for the regional ITV servers, including the ITV server on the Secret Internet Protocol Router Network (SIPRNET). PM-AIT is responsible for ensuring that ITV system performance and information assurance requirements are in accordance with DoD 8500.1, Information Assurance, and DoD 8500.2, Information Assurance (IA) Implementation. The Non-classified Internet Protocol Router Network (NIPRNET)-based ITV servers must be interoperable with GTN, GTN 21, Joint Total Asset Visibility, and Integrated Data Environment, and other DoD logistics systems as determined by the PM-AIT Office and the user representative(s). The SIPRNET-based ITV server must interoperate with the Global Combat Support System, Global Command and Control System, and other classified systems as determined by PM-AIT and the User Representative(s). PM-AIT is responsible for maintaining the accreditation and net worthiness certification of all ITV servers.

Wireless Encryption Requirements   Per the DoD Wireless Policy (DoD 8100.2), encryption requirements do not apply to the detection segment of a personal electronic device (PED), e.g., the laser used in optical storage media; between a bar code and a scanner head; or Radio Frequency (RF) energy between RF identification tags, both active and passive, and the reader/interrogator.

Frequency Spectrum Management   PM-AIT office will continue to assist DoD Components in frequency management issues related to active RFID tags and equipment purchased under the DoD RFID contracts by PM-AIT.

   RFID tags that meet the technical specifications of 47 CFR 15 of the FCC's Rules and Regulations for Non-Licensed Devices, i.e., Part 15, must accept and may not cause electromagnetic interference to any other federal or civil RF device. 47 CFR 15 only applies to use of these devices within CONUS and other US Possessions. DoD components will forward requests for frequency allocation approval via command channels to the cognizant military frequency management office to ensure that RFID tags comply with U.S. National and OCONUS host-nation spectrum management policies. RFID tags and infrastructure may require electromagnetic compatibility analysis to quantify the mutual effects of RFID devices within all intended operational environments, e.g., Hazards of Electromagnetic Radiation to Ordnance (HERO) and Hazards of Electromagnetic Radiation to Fuel (HERF). (References: International

Telecommunications Union (ITU) Radio Regulations (Article 5); National Telecommunications and Information Administration (NTIA) Manual of Regulations and Procedures for Federal Radio Frequency Management; DoD Directive 3222.3, Department of Defense Electromagnetic Compatibility Program, 20 Aug 1990; DoD Directive 4650.1, Policy for Management and Use of the Electromagnetic Spectrum, 8 Jun 04).

BUSINESS RULES FOR PASSIVE RFID TECHNOLOGY IN THE DOD SUPPLY CHAIN

Overview

Passive Radio Frequency Identification (RFID) tags reflect energy from the reader/interrogator or receive and temporarily store a small amount of energy from the reader/interrogator signal in order to generate the tag response. Passive RFID requires strong RF signals from the reader/interrogator, while the RF signal strength returned from the tag is constrained to low levels by the limited energy. This low signal strength equates to a shorter range for passive tags than for active tags. The DoD approved frequency range for passive RFID implementation is UHF 860–960 MHz.

The DoD Logistics Automatic Identification Technology (LOG-AIT) Office is the DoD focal point for coordinating overarching guidance for the use of AIT within DoD. The Program Executive Office, Enterprise Information Systems (PEG EIS), Product Manager-Automatic Identification Technology (PM-AIT) Office is the DoD procurement activity for AIT equipment (to include RFID equipment and infrastructure) and will establish a standing contract for equipment installation and maintenance. Beginning in FY 2007, only RFID capable AIT peripherals (e.g., optical scanners and printers used for shipping labels) will be acquired when those peripherals support RFID-capable business processes. Beginning in FY 2007, logistics automated information systems (AISs) involved in receiving, shipping, and inventory management will use RFID to perform business transactions, where appropriate. AIS funding will hinge on compliance with this policy. The Defense Logistics Board (DLB) will review these requirements prior to FY 2007 implementation.

Passive RFID Business Rules

The following prescribes the business rules for the application of passive Rfrn technology at the case, pallet, and item packaging (unit pack) for Unique Identification (urn) items on shipments to and within DoD. These rules are in addition to the urn requirement for data element identification of DoD tangible assets using 2D data matrix symbology marking on the item itself. To facilitate the use of RFrn events as transactions of record, the DoD has embraced the use of Electronic Product Code™ (EPC) tag data constructs, as well as DoD tag data constructs, in a supporting DoD data environment. As the available EPC technology matures, the intent is to expand the use of passive RFrn applications to encompass individual item tagging.

DoD RFID Definitions   The following definitions apply to passive RFID technology and tags in support of the DoD requirement to mark/tag materiel shipments to DoD activities in accordance with this policy:

EPC Technology: Passive RFID technology (readers, tags, etc.) that is built to the most current published EPCglobal™ Class O and Class 1 specifications and that meets interoperability test requirements as prescribed by EPCglobal™. EPC Technology will include Ultra High Frequency Generation 2 (UHF Gen 2) when this specification is approved and published by EPCglobal™.

Unit Pack: A MIL-STD-129 defined unit pack, specifically, the first tie, wrap, or container applied to a single item, or to a group of items, of a single stock number, preserved or unpreserved, which constitutes a complete or identifiable package.

Case: Consists of either an exterior container within a palletized unit load or an individual shipping container.

Exterior Container: A MIL-STD-129 defined container, bundle, or assembly that is sufficient by reason of material, design, and construction to protect unit packs and intermediate containers and their contents during shipment and storage. It can be a unit pack or a container with a combination of unit packs or intermediate containers. An exterior container may or may not be used as a shipping container.

Shipping Container: A MIL-STD-129 defined exterior container which meets carrier regulations and is of sufficient strength, by reason of material, design, and construction, to be shipped safely without further packing (e.g., wooden boxes or crates, fiber and metal drums, and corrugated and solid fiberboard boxes).

Pallet (palletized unit load): A MIL-STD-129 defined quantity of items, packed or unpacked, arranged on a pallet in a specified manner and secured, strapped, or fastened on the pallet so that the whole palletized load is handled as a single unit. A palletized or skidded load is not considered to be a shipping container.

Case, Palletized Unit Load, UID Item Packaging Tagging/Marking   DoD sites where materiel is associated into cases or pallets will tag the materiel and supplies at that site with an appropriate passive RFrn tag prior to further trans-shipment to follow-on consignees. The Defense Logistics Agency has committed to enabling the strategic distribution centers at Defense Distribution San Joaquin, CA (DDJC) and Defense Distribution Susquehanna, PA (DDSP) with passive RFrn capability by January 1, 2005.

Per the schedule outlined in the DOD attachment, case, pallet, and item packaging (unit pack) for Unique Identification (Urn) items will be tagged at the point of origin (including vendors) with passive RFrn tags, with some exceptions for the bulk commodities. If the unit pack for urn items is also the case, only one RFrn tag will be attached to the container.

Bulk Commodities Not Included   The following bulk commodities are defined as those that are shipped in rail tank cars, tanker trucks, trailers, other bulk wheeled conveyances, or pipelines.

- Sand
- Gravel
- Bulk liquids (water, chemicals, or petroleum products)
- Ready-mix concrete or similar construction materials
- Coal or combustibles such as firewood, agricultural products—seeds, grains, animal feeds, and the like

Contract/Solicitation Requirements   Per the schedule outlined in the DoD policy statement, new solicitations for materiel issued after October 1, 2004, for delivery after January 1, 2005, will contain a requirement for passive RFrn tagging at the case (exterior container within a palletized unit load or shipping container), pallet (palletized unit load), and the urn item packaging level of shipment in accordance with the appropriate interim/final Defense Federal Acquisition Regulation Supplement (DF ARS) Rule/Clause or MIL-STD-129 as appropriate.

Passive UHF RFID Tag Specifications   The DoD approved frequency range for the tags is 860–960 MHz with a minimum read range of three (3) meters. As the EPC UHF Gen 2 tag specification is distributed and quantities of UHF Gen 2 items are available for widespread use DoD shall adopt the EPC UHF Gen 2 tags.

The tags will be utilized for initial shipments from suppliers in compliance with appropriate contractual requirements to tag items shipped to DoD receiving points commencing January 1, 2005.

Since the UHF Gen 2 EPC technology is now approved, the DoD will establish firm tag acceptance expiration dates (sunset dates) for EPC Version 1 (class 0 and 1) tags and will now accept only UHF Gen 2 EPC tags. The DoD goal is to migrate to use of an open standard UHF Gen 2 EPC tag, Class 1 or higher, that will support DoD end-to-end supply chain integration.

As outlined below, suppliers to DoD must encode an approved tag using either a DoD tag data construct or an EPC tag data construct. Suppliers that choose to employ the DoD tag construct will use the Commercial and Government Entity (CAGE) code previously assigned to them and encode the tags per the rules that follow. Suppliers that are EPCglobal™ subscribers and possess a unique EPC manager number may choose to use the EPC tag data construct to encode tags per the rules that follow. Suppliers must ensure that each tag identification is unique.

Passive UHF RFID Tag Specifications

Passive UHF RFID Tag Data Structure Requirements Suppliers shipping to DoD—EPCglobal™ Subscribers using an EPCglobal tag™ data construct layout for 64 Bit EPCglobal™ Data Constructs or Layout for 96 Bit EPCglobal™ Data Constructs.

Sample binary encoding of the fields of a 64 bit Class 1 tag on a case shipped from DoD supplier. Complete Content String of the Above Encoded Sample Tag is provided in the complete version of the DoD policy statement.

Passive UHF RFID Tag Data Structure Requirements DoD initial implementations used currently available 64-bit tags but should transition to 96-bit tags as soon as practicable.

Electronic Data Interchange (EDI) Information

To effectively utilize RFID events to generate transactions of record in DoD logistics systems, RFID tag data with the associated material information must be resident in the DoD data environment so that information systems can access this data at each RFID event (i.e., tag read).

The DoD will require commercial suppliers to provide standard Ship Notice/Manifest Transaction Set (856) transactions in accordance with the Federal Implementation Convention (IC) via approved electronic transmission methods (EDI, web-based, or user defined format) for all shipments in accordance with the applicable DF ARS Rule via Wide Area Workflow (WAWF). Internal DoD sites/locations and shippers will use the EDI IC 856S or 856A, as applicable.

The transaction sets enable the sender to describe the contents and configuration of a shipment in various levels of detail and provide an ordered flexibility to convey information. The Federal IC 856 and DoD IC 856S and 856A transaction sets will be modified by the appropriate DoD controlling agencies to ensure the transactions can be used to list the contents for each piece of a shipment of goods as well as additional information relating to the shipment such as: order information, product description to include the item count in the shipment piece and item UID information, physical characteristics, type of packaging to include container nesting levels within the shipment, marking to include the shipment piece number and RFID tracking number, carrier information, and configuration of goods within the transportation equipment.

The DoD will also accept the submission of web-based ASN transactions as well as User-Defined-Format (UDF) ASN files. The following required ASN transactions will facilitate this use of RFID events.

RFID Funding

The cost of implementing and operating RFID technology is considered a normal cost of transportation and logistics and as such should be funded through routine Operations and Maintenance, Working Capital Fund, or Capital Investment processes. It is the responsibility of the DoD activity at which cases or palletized unit loads are built to procure and operate sufficient quantities of passive RFID equipment (interrogators/readers, write stations, tags, etc.) to support required operations. It is the responsibility of the activity at which cases or palletized unit loads are received (i.e., activity where the "supply" receipt is processed) to procure and operate sufficient quantities of passive RFID equipment (interrogators/readers) to support receiving operations. Working Capital Fund activities providing this support will use the most current DoD guidance in determining whether operating cost authority or capital investment program authority will be used to procure the required RFID equipment.

DoD Purchase Card Transactions

Per current DoD regulations, DoD Purchase Cards may be used to acquire items on existing government contracts as well as acquire items directly from suppliers that are not on a specific government contract. If the DoD Purchase Card is used to acquire items that are on a government contract that includes a requirement for RFID tagging of material per the appropriate DF ARS Rule, any items purchased via the DoD Purchase Card shall be RFID tagged in accordance with this policy. This policy does not apply to items acquired via a DoD Purchase Card that are not on a government contract. If DoD customers desire the inclusion of a passive RFID tag on shipments for these type purchases, this requirement must be specifically requested of the shipping supplier/vendor and the shipment must be accompanied by an appropriate ASN containing the shipment information associated to the appropriate RFID tag.

Wireless Encryption Requirements

Per the DoD Wireless Policy (DoDD 8100.2), encryption requirements do not apply to the detection segment of a personal electronic device (PED), e.g., the laser used in optical storage media; between a bar code and a scanner head; or Radio Frequency (RF) energy between RF identification tags, both active and passive, and the reader/interrogator.

Frequency Spectrum Management

RFID tags that meet the technical specifications of 47 CFR 15 of the FCC's Rules and Regulations for Non-Licensed Devices, i.e., Part 15, must accept and may not cause electromagnetic interference to any other federal or civil

RF device. 47 CFR 15 only applies to use of these devices within CONUS and other U.S. Possessions. DoD components will forward requests for frequency allocation approval via command channels to the cognizant military frequency management office to ensure that RFID tags comply with U.S. national and OCONUS host-nation spectrum management policies. RFID tags and infrastructure may require electromagnetic compatibility analysis to quantify the mutual effects of RFID devices within all intended operational environments, e.g., Hazards of Electromagnetic Radiation to Ordnance (HERO) and Hazards of Electromagnetic Radiation to Fuel (HERF). (References: International Telecommunications Union (ITU) Radio Regulations (Article 5); National Telecommunications and Information Administration (NTIA) Manual of Regulations and Procedures for Federal Radio Frequency Management; DoD Directive 3222.3, Department of Defense Electromagnetic Compatibility Program, 20 Aug 1990; DoD Directive 4650.1, Policy for Management and Use of the Electromagnetic Spectrum, 8 Jun 04).

SUPPLIER IMPLEMENTATION PLAN

Overview

Considering the volume of contracts and Department has developed a plan for passive warfighting customer. This implementation distribution functions within the Defense facilities, and strategic aerial ports. The variety of commodities managed, the RFID tagging that delivers best value to the plan provides a roadmap that targets critical Distribution Depots, depot maintenance.

Suppliers Shipping to DoD

Per the schedule outlined in this attachment, case, pallet, and item packaging (unit pack) for Unique Identification (urn) items will be tagged at the point of origin (manufacturer/vendor) with passive RFID tags, except for the bulk commodities as defined in the policy statement. If the unit pack is also the case, only one RFID tag will be attached to the container. Shipments of goods and materials will be phased in by procurement methods, classes/commodities, location, and layers of packaging for passive RFID.

Commencing January 1, 2005

All individual Cases plus All Cases packaged within Palletized Unit Loads plus all Palletized Unit Loads, will be tagged for the following commodities:

- Packaged Operational Rations (subclass of Class I)
- Clothing, Individual Equipment, Tools (Class ll)

- Personal Demand Items (Class VI)
- Weapon System Repair Parts and Components (Class IX)

When these commodities are being shipped to the following locations:

- Defense Distribution Depot, Susquehanna, PA (DDSP)
- Defense Distribution Depot, San Joaquin, CA (DDJC)

After January 1, 2006

All individual Cases plus All Cases packaged within Palletized Unit Loads plus all Palletized Unit Loads, will be tagged for the above commodities in addition to the following classes/commodities to be phased in pending appropriate safety certifications.

- Subsistence and Comfort Items (Class I)
- Packaged Petroleum, Lubricants, Oils, Preservatives, Chemicals, Additives (Class llIP)
- Construction and Barrier Material (Class IV)
- Ammunition of all types (Class V)
- Major End Items (Class VII)
- Pharmaceuticals and Medical Materials (Class VIII)

Item Packaging for UID items will be tagged if the packaging is the case or exterior of a palletized unit load.

Commencing January 1, 2007

RFID tagging will be required for all DoD manufacturers and suppliers who have new contracts, issued with the appropriate contract clause, according to the following implementation guidelines:

All classes of supply that will require RFID tags on all individual cases, all cases within palletized unit loads, all pallets, and all unit packs for unique identification (UI).

RFID tagging will be required on commodities that will be tagged which are shipped DoD location that has been instrumented.

DoD RFID Status

The Department of Defense remains committed (as noted in their RFID web page updated on September 26, 2006) to the implementation of Radio Frequency Identification (RFID) technology as outlined in their July 30, 2004 policy memorandum. Since the publication of this initial policy memorandum, ongoing technology developments, updated IT investment strategies, and business process improvements within the DoD have clarified passive RFID requirements within the Department. The DoD July 30, 2004 RFID Policy stated that passive RFID tagging by DoD suppliers would apply to all locations worldwide. The term "all locations" in the July 30, 2004 policy refers to all major receiving locations across the world. The DoD is investing in appropriate passive RFID infrastructure in all locations that are deemed major receiving locations; the majority of those locations are already called out in the current DFARS clause. The DoD requirement will expand to tactical locations as those locations become RFID-enabled. The DoD will not require suppliers to apply passive RFID tags to the packaging of UID items during the 2007 calendar year. The Department will continue to evaluate the appropriate time frame to begin tagging at the packaging level for UID items and will promulgate this requirement in advance of future issuances.

The Acting Under Secretary of Defense for Acquisition, Technology, and Logistics signed a memorandum outlining policy for the use of RFID within the Department of Defense (DoD). The strategy calls for taking maximum advantage of the inherent life-cycle asset management efficiencies that can be realized with integration of RFID throughout DoD.

Leveraging this technology to improve our ability to get the customer the right materiel, at the right time, and in the right condition is a critical part of our End-to-End Warfighter Support initiative.

The new policy addresses two general types of RFID tags: (1) active, which contains an internal power source, enabling the tag to hold more data and has a longer "read" range and (2) passive, which does not contain any power source, holds a minimum of data and has a shorter "read" range.

- The policy directs the adoption of specific business rules for the active, high data capacity RFID currently used in the DoD operational environment to ensure continued support for ongoing Combatant Commander in-transit visibility requirements and operations.
- The policy states that DoD will be an early adopter of innovative, passive RFID technology that leverages the Electronic Product Code (EPC) and compatible RFID tags. The policy will require suppliers to put passive RFID tags on lowest possible piece part/case/pallet packaging once the supplier's contract contains language regarding the requirement.

LIST OF ACRONYMS

ADS	Applied Digital Solutions
AIDC	Automatic Identification and Data Capture Technology
ANSI	American National Standards Institute
ASCII	American Standard Code for Information Exchange
ASK	Amplitude Shift Keying
BSI	British Standards Institute
CD-ROM	Compact Disc, Read-Only Memory
CDC	California Department of Corrections
CDRW	Compact Disc, Read/Write Memory
CEPT	European Conference of Postal and Telecommunications Administrations
CPG	Compliance Policy Guide
CSI	Container Security Initiative
CSI	Crime Scene Investigation
DLB	Defense Logistic Board
DNS	Domain Name Server
DoD	Department of Defense
EAN	European Article Numbering System
EAS	Electronic Article Surveillance
EDI	Electric Data Interchange
ECC	European Communications Committee
EFF	Electronic Frontiers Foundation

RFID-A Guide to Radio Frequency Identification, by V. Daniel Hunt, Albert Puglia, and Mike Puglia
Copyright © 2007 by Technology Research Corporation

EPC	Electronic Product Code
ERO	European Radiocommunications Office
ETSI	European Telecommunications Standards Institute
FSK	Frequency Shift Keying
GPS	Global Positioning System
GTIN	Global Trade and Identification Number
GTN	Global Transportation Network
HF	High Frequency
IEC	International Electro-technical Commission
ISM	International Scientific and Medical Band
ISO	International Organization for Standardization
IT	Information Technology
ITU	International Telecommunications Union
ITV	In-Transit Visibility
LAN	Local Area Network
LF	Low Frequency
MfrTagID	Manufacturers Tag ID
MPHPT	Ministry of Public Management, Home Affairs, Posts and Telecommunications
OCR	Optical Character Recognition
ONS	Object Naming Service
OOK	On-Off Keying
PDMA	Prescription Drug Marketing Act
PML	Product Markup Language
POS	Point of Sale
PSK	Phase Shift Keying
PTU	Personal Tracking Unit
RFID	Radio Frequency Identification
RF	Radio Frequency
RFrf	DoD Radio Frequency Terminology
RO	Read-Only Memory
ROA	Return on Assets
ROI	Return on Investment
RW	Read/Write Memory
SCM	Supply Chain Management
SST	Smart and Secure Tradelanes
TRC	Technology Research Corporation
UCC	Uniform Code Council
UHF	Ultra-high Frequency
UPC	Universal Product Code
WORM	Write-Once/Read-Many Memory

GLOSSARY

An edited glossary of RFID terms is included in this book based on the material reproduced with the permission of the Association for Automatic Identification and Mobility and Industry Usage. The basic original source was the AIM Inc. White Paper, Document Version: 1.2, 2001-08-23. This original material was reprinted with the permission of AIM Inc. Copyright © AIM, Inc.; www.aimglobal.org: www.rfid.org.

Active Tag Colloquial term for a radio frequency (RFID) transponder powered partly or completely by a battery. Batteries may be replaceable or sealed within the device (the term *unitized active tag* is sometimes used when the battery is sealed in the tag).

Active Transponder A battery-powered data-carrying device that reacts to a specific, reader-produced, inductively coupled or radiated electromagnetic field, by delivering a data-modulated radio frequency response.

Addressability The ability to address bits, fields, pages, files, or other defined areas of memory within a radio frequency tag.

Air Interface The conductor-free medium, usually air, between a transponder and the reader/interrogator through which data communication is achieved by means of a modulated inductive or propagated electromagnetic field.

AIM (Automatic Identification Manufacturers)—Generic abbreviation for Automatic Identification Manufacturer Trade Associations, including AIM Inc., AIM UK, AIM Germany, etc.

RFID-A Guide to Radio Frequency Identification, by V. Daniel Hunt, Albert Puglia, and Mike Puglia
Copyright © 2007 by Technology Research Corporation

Alignment A term to express the orientation of a transponder, relative to the reader/interrogator antenna. Alignment can influence the degree of coupling between transponder and reader, separation being a further influence.

Alphanumeric Strictly data comprising both alphabetical and numeric characters. For example, A1234C9 as an alphanumeric string. The term is often used to include other printable characters such as punctuation marks.

Amplitude Modulation (AM) Representation of data or signal states by the amplitude of a fixed frequency sinusoidal carrier wave. Where data is in binary form the modulation involves two levels of amplitude and is referred to as amplitude shift keying (ASK).

Amplitude Shift Keying (ASK) Representation of binary data states, 0 and 1, by the amplitude of a fixed frequency sinusoidal carrier wave. Where the amplitudes are determined by the carrier being switched on and off, the process is known as On-Off Keying (OOK).

Antenna A conductive structure specifically designed to couple or radiate electromagnetic energy. In a driven mode, the structure is a transmitter antenna. In receiver mode, the structure is a receiver antenna. Antenna structures, often encountered in radio frequency identification systems, may be used to both transmit and receive electromagnetic energy, particularly data modulated electromagnetic energy. See also dipole.

Anti-clash A term describing a facility for avoiding contention at the reader/interrogator receiver for responses arising from transponders simultaneously present within the read or interrogation zone of a radio frequency identification system and competing for attention at the same time.

ASCII (American Standard Code for Information Interchange) A binary code comprising 128 alphanumeric and control characters, each encoded with 7 bits.

ASN.1 (Abstract Syntax Notation 1) A syntax language for communicating processes, including transfer syntax or rules for converting variables, commands, and requests into forms that are hardware independent.

Asynchronous Transmission A method of data transmission that does not require timing or clocking information in addition to data. Transmission is achieved by receiver reference to start and stop bits positioned at the beginning and end of each character or blocks of characters. A variable time interval can exist between characters or blocks of characters.

Awake The condition of a transponder when it is able to respond to interrogation.

Backscatter Modulation A process whereby a transponder responds to a reader/interrogation signal or field by modulating and re-radiating or transmitting the response signal at the same carrier frequency.

Batch Reading The process or capability of a radio frequency identification reader/interrogator to read a number of transponders present within the system's interrogation zone at the same time. Alternative term for *multiple reading*.

Bandwidth The range or band of frequencies, defined within the electromagnetic spectrum, that a system is capable of receiving or delivering.

Baud A unit of signaling or transmission speed representing the number of signaling events per unit time. When the signal event is a single bit, binary state representation, the baud is equivalent to the bit rate, expressed in bits per second (bps).

BCC (Block Check Character) A parity error checking character added to data for the purposes of detecting transmission errors.

BER (Bit Error Rate) The ratio of the number of bits received in error to the total of bits transmitted.

Binary Number System A term used to describe the capability of two way communication. A column-placing numbering system in which numbers are expressed as powers of 2 (. . . 23 22 21 20) using the digits 0 and 1 to distinguish the weighting of powers to represent the number concerned. For example, $2510 = 0 \times 25\ 1 \times 24\ 1 \times 23\ 0 \times 22\ 0 \times 21\ 1 \times 20 = 011001$.

Binary Coded Decimal (BCD) Representation of decimal numbers in binary form using a group of four bits to represent an individual digit (0–9). For example, 0011 1000 = 3810.

Bit Abbreviation for binary digit. A single element (0, 1) in a binary number.

Bit Rate Rate at which bits are electronically communicated, measured in bits.

Bit Error Rate (BER) The ratio of the number of bits received that are found to be in error to the total number of bits transmitted.

Bi-phase Coding A generic term for line or channel encoding schemes of the Manchester coding type in which each bit in the source code is replaced by two bits in the derived format.

Block Code Error detection codes having a fixed length code format, wherein k message bits are accompanied by c parity bits to form an n-bit block code $(n = k + c)$

Bulk Commodities These items shall not be tagged in accordance with passive RFID tagging requirements. Bulk commodities are products carried or shipped in rail tank cars; tanker trucks; other bulk, wheeled conveyances; or pipelines.

Byte A group of bits, usually eight, used to represent characters in a binary processing system.

Capacity–Data A measure of the data, expressed in bits or bytes, that can be stored in a transponder. The measure may relate simply to the bits that are accessible to the user or to the total assembly of bits, including data identifier and error control bits.

Capacity–Channel   A measure of the transmission capability of a communication channel expressed in bits-1 and related to channel bandwidth and signal to noise ratio by the Shannon equation: Capacity, $C = B \log^2 (1 + S/N)$, where B is the bandwidth and S/N the signal to noise ratio.

Capture Field/Area/Zone (also Interrogation Zone/Area/Volume)   The region of the electromagnetic field, determined by the reader/interrogator antenna, in which the transponders are signaled to deliver a response.

Carrier   Abbreviated term for *Carrier Frequency*.

Carrier Frequency   The frequency used to carry data by appropriate modulation of the carrier waveform, typically in a radio frequency identification system, by amplitude shift keying (ASK), frequency shift keying (FSK), phase shift keying (PSK), or associated variants.

Case   Consists of either an exterior container within a palletized unit load or an individual shipping container.

Channel   A medium or medium-associated allocation, such as carrier frequency, for electronic communication.

Channel Encoding   The application of coding schemes to facilitate effective channel transmission of the source encoded data.

Channel Decoding   The process of operating upon a received transmission to separate the source-encoded data from the channel encoded form.

Character Set   A set of characters assembled to satisfy a general or application requirement.

Checksum   A summation of check digits used to determine if an error has occurred in the transmission of data.

Chip   In data communication terms, the smallest duration of a pseudo-random code sequence used in spread spectrum communication systems.

Chipping   The insertion of a RFID tag under the skin of an individual for identification or tracking applications.

Clocking Information   Timing signals or pulses used to synchronize the transfer of data from a source to a host destination.

Closed Systems   Within the context of radio frequency identification, they are systems in which data handling, including capture, storage, and communication are under the control of the organization to which the system belongs.

Code Plate   An alternative colloquial term for *transponder* or *tag*.

Collision   A term to denote an event in which two or more data communication sources compete for attention at the same time and cause a clash of data, inseparable without some means of anti-collision or contention management.

Collision Avoidance   A means of avoiding collisions or clashes of data from different sources competing for attention at the same time.

Compatibility The condition that exists between devices or systems that exhibit equivalent functionality, interface features and performance to allow one to be exchanged for another, without alteration, and achieve the same operational service. An alternative term for *Interchangeability*.

Concatenation The facility to link together specific items of data, held in data carriers, to form a single file or field of data.

Continuous Reporting A mode of reader/interrogator operation wherein the identification of a transponder is reported or communicated continuously while the transponder remains within the interrogation field.

Continuous Wave Modulation A data modulation scheme in which the data is represented by the carrier signal being switched on and off. The scheme is identical to amplitude shift keying (ASK) with 100% depth of modulation—known as on-off keying (OOK).

Control Characters Characters within a character set which are used to denote a particular control function, such as new line, shift and print control.

Concentrator A means of connecting a number of data communication devices and concentrating packets of data at a local point before onward transmission on a single link to a central data processor or information management system. In contrast to multiplexors concentrators usually have a buffering capability to "queue" inputs that would otherwise exceed transmission capacity.

Contention (Clash) Term denoting simultaneous transponder responses capable of causing potential confusion, and misreading, within a reader/interrogator system unequipped with anti-contention facilities.

Corruption–Data In data terms, the manifestations of errors within a transmitted data stream due to noise, interference, or distortion.

Data Representations, in the form of numbers and characters for example, to which meaning may be ascribed.

Data Rate In a radio frequency identification system, the rate at which data is communicated between transponder and the reader/interrogator, expressed in baud, bits-1 or bytes-1.

Data Field A defined area of memory assigned to a particular item or items of data.

Data Field Protection The facility to control access to and operations upon items or fields of data stored within the transponder.

Data Identifier A specific character, or string of characters, that denotes the nature or intended use of the data that follows.

Data Transfer The process of transferring data from a data holding source to a destination.

Demodulation Process of recovering channel encoded data from a modulated carrier waveform.

De-tuning   The reduction in performance of transponders and readers/interrogators caused by the close proximity of metal influencing the resonance of an electronic tuned circuit.

Dipole (Antenna)   A fundamental form of antenna, comprising a single conductor of length approximately equal to half the wavelength of the carrier wave. Provides the basis for a range of other more complex forms of antenna.

Directivity–antenna   The ability of an antenna to concentrate radiated energy in a preferred direction, when considered in a transmitter mode. Alternatively, the ability to reject signals that are off-axis to the normal of the antenna, when considered in the receiver mode. May be expressed as a ratio of power radiated per unit solid angle in a defined direction to the total power radiated by the antenna.

Direct Sequence Spread   A category of spread spectrum modulation in which the source base-band bit stream is multiplied by a fast pseudo random binary sequence to produce a signal that exhibits broad-band characteristics. Alternatively, the pseudo random sequence and its inverse are used to represent logic 1 and 0.

Dispersion–pulse   The spread in duration and form experienced by a pulse in transmission through a communication channel.

Distortion   Any disturbance that causes an unwarranted change in the form or intelligibility of a signal. The distortion exhibits a noise-like effect that can be quantified as the ratio of the magnitude of the distortion component to the magnitude of the undistorted signal, usually expressed as a percentage.

Downlink   Term which defines the direction of communications as being from reader/interrogator to transponder.

Effective Aperture   A term denoting the reception capability of a practical antenna expressed as the product of actual aperture and antenna efficiency.

Efficiency–antenna   Two components distinguishable, radiation efficiency and aperture efficiency. Radiation efficiency is expressed as the ratio of total power radiated by the antenna to total power accepted by the antenna from source, for the transmission mode. Aperture efficiency is expressed as the ratio of effective antenna area to the real area of the antenna.

Electromagnetic Coupling   A process of transferring modulated data or energy from one system component to another, reader to transponder, for example, by means of an electromagnetic field.

Electromagnetic Field   The spatial and temporal manifestation of an electromagnetic source in which magnetic and electric components of intensity can be distinguished and plotted as contours, like contour lines on a map, the planes of the electric and magnetic contours being at right angles to one another. Where the source is varying in time, so too the field components vary with time. Where the source launches an electromagnetic wave, the field may be considered to be propagating.

Electromagnetic Spectrum   The range or continuum of electromagnetic radiation, characterized in terms of frequency or wavelength.

Electromagnetic Wave   A sinusoidal wave in which electric E and magnetic H components or vectors can be distinguished at right angles to one another, and propagating in a direction that is at right angles to both the E and H vectors. The energy contained within the wave also propagates in the direction at right angles to the E and H vectors. The power delivered in the wave is the vector product of E and H (Pointing Vector).

Electronic Data Interchange   Communication of a data message, or messages, automatically between computers or information management systems, usually for the purposes of business transactions.

Electromagnetic Waves   Electromagnetic waves are characterized by two field components, a magnetic (H) component and an electric (E) component. These components are mutually perpendicular to each other and to the direction of propagation.

Electronic Data Transfer   The transfer of data by electronic communication means from one data handling system to another.

Electronic Label   An alternative colloquial term for a transponder.

Encryption of Data   A means of securing data, often applied to a plain or clear text, by converting it to a form that is unintelligible in the absence of an appropriate decryption key.

Environmental Parameters   Parameters, such as temperature, pressure, humidity, and noise, that can have a bearing or impact upon system performance.

EPCglobal   EPCglobal is a joint venture between European Article Numbering International (EAN) and the Uniform Code Council (UCC) (GS1) and is leading an effort to create global standards for RFID use.

EPCglobal Inc is an open, worldwide, not-for-profit consortium of supply chain partners working to drive global adoption of the EPCglobal Network™. Using Electronic Product Code™ (EPC) and Radio Frequency Identification (RFID) technologies, the EPCglobal Network will provide for immediate, automatic and accurate identification of any item in the supply chain of any company, in any industry, anywhere in the world. For more information about EPCglobal, visit http://www.epcglobalinc.org.

EPC Technology   Passive RFID technology (readers, tags, etc.) that is built to the most current published EPCglobal™ Class O and Class 1 specifications and that meets interoperability test requirements as prescribed by EPCglobal™.

Error   In digital data terms, a result of capture, storage, processing, or communication of data in which a bit or bits assume the wrong values, or bits are missing from a data stream.

Error Burst   A group of bits in which two successive erroneous bits are always separated by less than a given number of correct bits.

Error Control   Collective term to accommodate error detection and correction schemes applied to handle errors arising within a data capture or handling system.

Error Detection   A term to denote a scheme or action to determine the presence of errors in a data stream.

Error Correction   A term to denote a scheme or action for correcting an error detected in a data stream.

Error Correcting Code (ECC)   Supplemental bits introduced or source encoded into a data stream to allow automatic correction of erroneous bits and/or derivation of missing bits, in accordance with a specific computational algorithm.

Error Correcting Mode   Mode defined for a data communication or handling process in which missing or erroneous bits are automatically corrected.

Error Correcting Protocol   The rules by which an error correcting mode operates.

Error Management   Techniques used to identify and/or correct errors within a data capture and handling system with the objective of assuring the accuracy of data presented to the system user.

ETSI (European Telecommunications Standards Institute)   The European standards organization responsible for standardization in telecommunications.

Exciter   The electronic circuits used to drive an antenna. The combination of exciter and antenna is often referred to as the transmitter or scanner.

Extended Binary Coded Data Interchange Code (EBCDIC)   An eight-bit binary code set, sometimes referred to as extended ASCII, wherein the 128 character set of ASCII are accommodated, together with other characters and control functions, making up a total set of 256 characters.

Exterior Container   A MIL-STD-129 defined container, bundle, or assembly that is sufficient by reason of material, design, and construction to protect unit packs and intermediate containers and their contents during shipment and storage. It can be a unit pack or a container with a combination of unit packs or intermediate containers. An exterior container may or may not be used as a shipping container.

Factory Programming   The entering of data into a transponder as part of the manufacturing process, resulting in a read-only tag.

False Activation   The result of a "foreign" or non-assigned transponder entering the interrogation zone of a radio frequency identification system and effecting a response, erroneous or otherwise.

Far Field   The region of an electromagnetic radiation field at a distance from the antenna in which the field distribution is unaffected by the antenna structure and the wave propagates as a plane wave.

Field of View   The zone surrounding a reader/interrogator in which the reader/interrogator is capable of communicating with a transponder.

Field Programming   Entry of data by an original equipment manufacturer (OEM) or user into a transponder by means of a proprietary programming system, usually undertaken before the device is attached to the item to be identified or accompanied. This facility is usually associated with write once read many (WORM) and read/write (RW) devices. The data entered into a transponder may be by a combination of factory and field programming.

Field Strength   The intensity of a field measured in units appropriate to the field concerned. Electric field strengths are measured in volts per meter and magnetic field strengths in amperes per meter.

File   A set of data stored within a computer, portable data terminal, or information management system.

Filler Character   A redundant character inserted into a data field simply to achieve a desired field length. Also known as a *pad character*.

Forward Link   Communications from reader/interrogator to transponder. Alternatively known as *downlink*.

Frequency   The number of cycles a periodic signal executes in unit time. Usually expressed in hertz (cycles per second) or appropriate weighted units such as kilohertz (kHz), megahertz (MHz), and gigahertz (GHz).

Frequency Hop Rate   The frequency at which a frequency hopping spread spectrum (FHSS) system moves between transmission frequencies. It is equal to the reciprocal of the dwell time at an FHSS center frequency.

Frequency Hop Sequence   A pseudo random binary sequence (PRBS) determining the hopping frequencies used in frequency hopping spread spectrum (FHSS) systems.

Frequency Hopping Spread Spectrum (FHSS)   A category of spread spectrum modulation in which each bit of data is divided into chips and each chip is represented by a different spectral component or tone in the spread spectrum band using a pseudo random sequence to assign tones. Modulated in this way, the transmissions hop from frequency to frequency within the band, requiring a receiver synchronized to the pseudo random chipping sequence to recover the data.

Frequency Modulation (FM)   Representation of data or signal states by using different transmission frequencies. Where data is in binary form the modulation constitutes two transmission frequencies and is referred to as *frequency shift keying* (FSK).

Frequency Shift Keying (FSK)   Representation of binary data by switching between two different transmission frequencies.

Full Duplex (FDX)   A channel communications protocol that allows a channel to transmit data in both directions at the same time. In RFID, the method of information exchange in which the information is communicated while the transceiver transmits the activation field.

Handshaking A protocol or sequence of signals for controlling the flow of data between devices, which can be hardware implemented or software implemented.

Half Duplex (HDX) A channel communications protocol that allows a channel to transmit data in both directions but not at the same time. In RFID, the method of information exchange in which the information is communicated after the transceiver has stopped transmitting the activation field.

Harmonics Multiples of a principal frequency, invariably exhibiting lower amplitudes. Harmonics can be generated as a result of circuit non-linearities associated with radio transmissions resulting in harmonic distortion.

Hexadecimal (Hex) A column placing method of representing data to the base of 16, using digits 0–9 and letters A to F for decimal values 10–15. For example, 1010 = A16 and 2210 = 6F16. Used as a convenient short-hand notation for representing 16- and 32-bit memory addresses.

ID Filter A software facility that compares a newly read identification (ID) with those within a database or set, with a view to establishing a match.

Impact Any influence upon a system, environmental or otherwise, that can influence its operational performance.

Incorrect Read The failure to read correctly all or part of the data set intended to be retrieved from a transponder during read or interrogation process. Alternative term for *misread*.

In-Field Reporting A mode of operation in which a reader/interrogator reports a transponder ID on entering the interrogation zone and then refrains from any further reports until a prescribed interval of time has elapsed.

In-Use Programming The ability to read from and write to a transponder while it is attached to the object or item for which it is being used.

Inductive coupling A process of transferring modulated data or energy from one system component to another, reader to transponder, for example, by means of a varying magnetic field.

Information–general Something that is meaningful. Data may be regarded as information once its meaning is revealed.

Information–theoretic A measure of the scarcity or probability of occurrence of an event, the more scarce the event, the more information conveyed.

Interface A physical or electrical interconnection between communicating devices. See also RS232, RS422, and RS485.

Interference Unwanted electromagnetic signals, where encountered within the environment of a radio frequency identification system, cause disturbance in its normal operation, possibly resulting in bit errors, and degrading system performance.

Interchangeability   The condition that exists between devices or systems that exhibit equivalent functionality, interface features, and performance to allow one to be exchanged for another, without alteration, and achieve the same operational service. An alternative term for *compatibility*.

Interoperability   The ability of systems, from different vendors, to execute bi-directional data exchange functions in a manner that allows them to operate effectively together.

Interrogation   The process of communicating with and reading a transponder.

Interrogator   A fixed or mobile data capture and identification device using a radio-frequency electromagnetic field to stimulate and effect a modulated data response from a transponder or group of transponders present in the interrogation zone. Often used as an alternative term to *reader*.

Interrogation Zone   The region in which a transponder or group of transponders can be effectively read by an associated radio frequency identification reader/interrogator.

Intersymbol Interference   Interference arising within a serial bit stream as a result of pulse dispersion and consequential overlapping pulse edges, leading possibly to decoding errors at the receiver.

Isotropic Source   An ideal electromagnetic source or radiator exhibiting a perfect spherical energy radiation pattern.

Lifetime   The period of time during which an item of equipment exists and functions according to specification. See also *Mean Time Between Failures* and *Mean Time to Repair*.

Manchester Coding   A bi-phase code format in which each bit in the source encoded form is represented by two bits in the derived or channel encoded form. The transformation rule ascribes 01 to represent 0 and 10 to represent 1.

Manufacturers Tag ID (MfrTagID)   A reference number that uniquely identifies the tag.

Mean Time Between Failures   The average or mean time interval between failures, often expressed as the reciprocal of the constant failure rate.

Mean Time to Repair   The length of time that a system is non-operational between failure and repair.

Memory   A means of storing data in electronic form. A variety of random access (RAM), read-only (ROM), write once–read many (WORM), and read/write (RW) memory devices can be distinguished.

Memory Modules   Colloquial term for a read/write or re-programmable transponder.

Misread   A condition that exists when the data retrieved by the reader/interrogator is different from the corresponding data within the transponder. Alternative term for *incorrect read*.

Modulation   A term to denote the process of superimposing (modulating) channel encoded data or signals onto a radio frequency carrier to enable the data to be effectively coupled or propagated across an air interface. Also used as an associative term for methods used to modulate carrier waves. Methods generally rely on the variation of key parameter values of amplitude, frequency, or phase. Digital modulation methods principally feature amplitude shift keying (ASK), frequency shift keying (FSK), phase shift keying (PSK), or variants. See also *Amplitude, Frequency and Phase Modulation, Amplitude Shift Keying, Frequency Shift Keying*, and *Phase Shift Keying*.

Modulation Index   The size of variation of the modulation parameter (amplitude, frequency, or phase) exhibited in the modulation waveform.

Multiple Reading   The process or capability of a radio frequency identification reader/interrogator to read a number of transponders present within the system's interrogation zone at the same time.

Multiplexor (Multiplexer)   A device for connecting a number of data communication channels and combining the separate channel signals into one composite stream for onward transmission through a single link to a central data processor or information management system. At its destination the multiplexed stream is de-multiplexed to separate the constituent signals. Multiplexors are similar to concentrators in many respects, a distinction being that concentrators usually have a buffering capability to "queue" inputs that would otherwise exceed transmission capacity.

Noise   Unwanted extraneous electromagnetic signals encountered within the environment, usually exhibiting random or wide band characteristics, and viewed as a possible source of errors through influence upon system performance.

Noise Immunity   A measure of the extent or capability of a system to operate effectively in the presence of noise.

Omnidirectional   A description of a transponder's ability to be read in any orientation.

On-Off Keying (OOK)   A special case of amplitude shift keying (ASK) in which the carrier is switched between full carrier amplitude and zero or absence of carrier amplitude, according to data value (1 or 0).

Open Systems   Within the context of radio frequency identification, they are systems in which data handling, including capture, storage, and communication, is determined by agreed standards, so allowing various and different users to operate without reference to a central control facility.

Orientation   The attitude of a transponder with respect to the antenna, expressed in three-dimensional angular terms, with range of variation expressed in terms of skew, pitch, and roll.

Orientation Sensitivity   The sensitivity of response for a transponder expressed as a function of angular variation or orientation.

Out of Field Reporting   A mode of operation in which the identification of a transponder is reported as or once the transponder leaves the reader interrogation zone.

Pallet (palletized unit load)   A MIL-STD-129 defined quantity of items, packed or unpacked, arranged on a pallet in a specified manner and secured, strapped, or fastened on the pallet so that the whole palletized load is handled as a single unit. A palletized or skidded load is not considered to be a shipping container.

Parity   A simple error detecting technique, used to detect data transmission errors, in which an extra bit (0 or 1) is added to each binary represented character to achieve an even number of 1 bits (even parity) or an odd number of 1 bits (odd parity). By checking the parity of the characters received a single errors can be detected. The same principle can be applied to blocks of binary data.

Passive Transponder (Tag)   A battery-free data carrying device that reacts to a specific, reader-produced, inductively coupled or radiated electromagnetic field, by delivering a data modulated radio frequency response. Having no internal power source, passive transponders derive the power they require to respond from the reader/interrogator's electromagnetic field.

Penetration   Term used to indicate the ability of electromagnetic waves to propagate into or through materials. Non-conducting materials are essentially transparent to electromagnetic waves, but absorption mechanisms, particularly at higher frequencies, reduce the amount of energy propagating through the material. Metals constitute good reflectors for freely propagating electromagnetic waves, with very little of an incident wave being able to propagate into the metal surface.

Phase Modulation (PM)   Representation of data or signal states by the phase of a fixed frequency sinusoidal carrier wave. Where data is in binary, form the modulation involves a phase difference of 180° between the binary states and is referred to as *Phase Shift Keying (PSK)*.

Phase Shift Keying (PSK)   Representation of binary data states, 0 and 1, by the phase of a fixed frequency sinusoidal carrier wave, a difference of 180° being used to represent the respective values.

Polar Field Diagram   A graphical representation of the electric or magnetic field intensity components of an electromagnetic field, expressed on a polar co-ordinate system. Typically used to illustrate the field characteristics of an antenna.

Polarization   The locus or path described by the electric field vector of an electromagnetic wave, with respect to time.

Polarization Summary   Polarization is a term that often arises in the literature and when considering radio frequency communication and RFID. The polarization of a propagating wave is determined by the locus or path described by the electric field vector with respect to time. If we ascribe an x, y, z co-ordinate system to a propagating wave as illustrated below, with the direction of propagation being in the z direction, the electric field vector, E, will be in the x, y plane. If E remains in the same orientation with respect to time, so that its locus describes a straight line, the wave is said to be linearly polarized. However, if the locus describes a circular motion with respect to time the wave is said to be circularly polarized. Where the locus describes an elliptical path the wave is said to be elliptically polarized.

Circular polarization is often used in communication systems since the orientation of the transmitting and receiving antenna is less important than it is with linearly polarized waves. The magnetic vector, H, always remains perpendicular to the E vector. Using an IEEE convention, a clockwise circular rotating wavefront approaching a receiver is defined as being left-hand circular (LHC) polarized.

Port Concentrator   A device that accepts the outputs from a number of data communication interfaces for onward transmission into a communications network.

Power Levels and Flux Density   The vector product of electric and magnetic field strengths within an electromagnetic wave, expressed as levels in watts and as a power flux density, measured at a distance from the source, in watts per square meter. Low power radio frequency transmissions are generally expressed in milli- or microwatts.

It is usual to express the levels and flux densities in terms of decibels, whereby the power level is referenced to an appropriate level, such as a watt or a milliwatt.

Programmability   The ability to enter data and to change data stored in a transponder.

Programmer   An electronic device for entering or changing (programming) data in a transponder, usually via a close proximity, inductively coupled data transfer link.

Programming   The act of entering or changing data stored in a transponder.

Projected Lifetime   The estimated lifetime for a transponder often expressed in terms of read and/or write cycles or, for active transponders, years, based upon battery life expectancy and, as appropriate, read/write activity.

Protocol   A set of rules governing a particular function, such as the flow of data/information in a communication system.

Proximity   Term often used to indicate closeness of one system component with respect to another, such as that of a transponder with respect to a reader.

Proximity Sensor   An electronic device that detects and signals the presence of a selected object. When used in association with a radio frequency identification system, the sensor is set up to sense the presence of a tagged or transponder carrying object when it enters the vicinity of the reader/interrogator so that the reader can then be activated to effect a read.

Pulse Dispersion   The spread in width or duration of a pulse during transmission through a practical transmission system, due to the influence of distributed reactive components

Radio Frequency Identification   An automatic identification and data capture system comprising one or more reader/interrogators and one or more transponders in which data transfer is achieved by means of suitably modulated inductive or radiating electromagnetic carriers.

Radio Frequency Tag   Alternative term for a *transponder*.

Range–Read   The maximum distance between the antenna of a reader/interrogator and a transponder over which the read function can be effectively performed. The distance will be influenced by orientation and angle with respect to the antenna, and possibly by environmental conditions.

Range–Programming   The maximum distance between the antenna of a reader/interrogator and a transponder over which a programming function can be effectively performed. Usually shorter than the read range, but may be influenced by orientation and angle with respect to the antenna, and possibly by environmental conditions.

Read   The process of retrieving data from a transponder and, as appropriate, the contention and error control management, and channel and source decoding required to recover and communicate the data entered at source.

Readability   The ability to retrieve data under specified conditions.

Reader/Interrogator   An electronic device for performing the process of retrieving data from a transponder and, as appropriate, the contention and error control management, and channel and source decoding required to recover and communicate the data entered at source.

Reader/Writer   The device may also interface with an integral display and/or provide a parallel or serial communications interface to a host computer or industrial controller.

Read Only   Term applied to a transponder in which the data is stored in an unchangeable manner and can therefore only be read.

Read Rate   The maximum rate at which data can be communicated between transponder and reader/interrogator, usually expressed in bits per second.

Read/Write   Applied to a radio frequency identification system, it is the ability both read data from a transponder and to change data (write process) using a suitable programming device.

Redundancy   In information terms, it describes the additional bits, such as those for error control or repeated data, over and above those required for transmitting the information message.

Reprogrammability   The ability to change the data content of a transponder using a suitable programming device.

RF Tag   Alternative, short-hand term for a transponder.

RS232 A common physical interface standard specified by the EIA for the interconnection of devices. The standard allows for a single device to be connected (point-to-point) at baud values up to 9,600 bps, at distances up to 15 meters. More recent implementations of the standard may allow higher baud values and greater distances.

RS422 A balanced interface standard similar to RS232, but using differential voltages across twisted pair cables. Exhibits greater noise immunity than RS232 and can be used to connect single or multiple devices to a master unit, at distances up to 3,000 meters.

RS485 An enhanced version of RS422, which permits multiple devices (typically 32) to be attached to a two wire bus at distances of over one kilometer.

SAW (Surface Acoustic Wave) Devices Devices using a transponder technology in which low power microwave signals are converted to ultrasonic waves by and on the surface of a piezoelectric crystal material forming the tag. Surface applied "finger" transducers determine the form and data content of the reflected return signal.

Scrambling The rearrangement or transposition of data to enhance security of stored data or the effectiveness of error control schemes.

Scanner The combination of antenna, transmitter (or exciter), and receiver into a single unit is often referred to as a scanner. With the addition of electronics to perform the necessary decoding and management functions to deliver the source data, the unit becomes a reader.

Screening The process of avoiding or minimizing electromagnetic interference by use of electromagnetic reflective and absorptive materials, suitably structured or positioned to reduce interaction between the source of potential interference and the circuit being protected.

Sensor An electronic device that senses a physical entity and delivers an electronic signal that can be used for control purposes.

Separation A term used to denote the operational distance between two transponders.

Shipping Contain A MIL-STD-129 defined exterior container which meets carrier regulations and is of sufficient strength, by reason of material, design, and construction, to be shipped safely without further packing (e.g., wooden boxes or crates, fiber and metal drums, and corrugated and solid fiberboard boxes).

Signal to Noise (S/N) The ratio of signal level to the level of noise present in a system, usually expressed in decibels.

Signal to Noise & Distortion The ratio of combined signal, noise, and distortion levels to the combined level of noise and distortion present in a system.

Sinusoidal Carrier A fundamental waveform, characterized by a single frequency and wavelength, used to carry data or information by modulating some feature of the waveform.

Source Decoding   The process of recovering the original or source data from a received source encoded bit stream.

Source Encoding   The process of operating upon original or source data to produce an encoded message for transmission.

Spectrum–Electromagnetic   The continuum of electromagnetic waves, distinguished by frequency components and bands that exhibit particular features or have been used for particular applications, including radio, microwave, ultraviolet, visual, infrared, X-rays and gamma rays.

Spectrum–Signal   Expression used to denote the make-up of a signal or waveform in terms of sinusoidal components of different frequency and phase relationship (spectral components).

Spectrum Mask   The maximum power density of a transmission expressed as a function of frequency.

Spurious Emissions   Usually denotes unwanted electromagnetic harmonics. Type approval testing includes measurement of harmonic emissions arising from the reader, to ensure they are within specified limits.

Spread Spectrum   Techniques for uniformly distributing or spreading the information content of a data carrying signal over a frequency range considerably larger than required for narrow band communication, allowing data to be recoverable under conditions of strong interference and noise.

SRD (Short Range Device)   A tag that is used at short range (less than 100 mm).

Synchronization   The process of controlling the transmission of data using a separate or derived clocking signal.

Synchronous Transmission   A method of data transmission that requires timing or clocking information in addition to data.

Tag   Colloquial term for a transponder. Commonly used and the term preferred by AIM for general usage.

Tolerance   The maximum permissible deviation of a system parameter value, caused by any system or environmental influence or impact. Usually expressed in parts per million (ppm).

Tolerances are specified for a number of radio frequency parameters, including carrier frequencies, sub-carriers, bit clocks, and symbol clocks.

Transceiver   A transmitter/receiver device used to both receive and transmit data.

Transmitter (Exciter)   An electronic device for launching an electromagnetic wave or delivering an electromagnetic field for the purpose of transmitting or communicating energy or modulated data/information. Often considered separately from the antenna, as the means whereby the antenna is energized. In this respect it is also referred to as an exciter.

Transponder   An electronic transmitter/responder, commonly referred to as a *tag*.

Unit Pack   A MIL-STD-129 defined unit pack, specifically, the first tie, wrap, or container applied to a single item, or to a group of items, of a single stock number, preserved or unpreserved, which constitutes a complete or identifiable package.

Unitized Active Tag   An active tag or transponder in which the batteries are replaceable or sealed within the device.

Uplink   Term that defines the direction of communications as being from transponder to reader/interrogator.

Vector   A quantitative component that exhibits magnitude, direction, and sense.

Verification   The process of assuring that an intended operation has been performed.

Write   The process of transferring data to a transponder, the internal actions of storing the data, which may also encompass the reading of data to verify the data content.

Write Rate   The rate at which data is transferred to a transponder and stored within the memory of the device and verified. The rate is usually expressed as the average number of bits or bytes per second over which the complete transfer is performed.

Write Once Read Many (WORM)   Distinguishing a transponder that can be part or totally programmed once by the user, and thereafter only read.

RFID VENDOR LIST

The vast number of companies interested in developing and marketing RFID products is changing daily. A snapshot of some of the RFID companies participating in the growth of RFID is presented in this section.

The *RFID Journal* has graciously permitted us to reproduce part of the extensive effort they make on a daily basis to provide current vendor information. Since any published book can only present this data in the form of a snapshot at the time created, we encourage you to update this type of information by referring directly to the *RFID Journal*.

This vendor list is included herein with the express permission to reprint by the **RFID Journal**. The original internet source is http://www.rfidjournal. com/article/findvendor.

We have provided below, many of the key vendors noted by the **RFID Journal**. In some cases, we deleted some vendors simply to conserve space and focus on the more significant entries.

3M

Based in St. Paul, Minnesota, the 3M company's RFID Tracking Solutions provides a tracking and management system for physical files that are stored in centralized file rooms or move through individual offices, workrooms or processing stations. Another product, 3M Library Systems, allows libraries to improve customer service and staff efficiency via RFID-based identification and tracking of library materials, simplified checkout and check-in, and automated sorting and inventory management.

A3 Technologies

A3 Technologies is a full-service data collection resource with experience in barcode, wireless, and RFID technologies. A3 develops system solutions with stand-alone barcode and RFID elements, or a mix of both. In RFID, A3 has experience with both passive and active tag technologies utilizing fixed-station and mobile readers. A3 implements passive solutions in the LF, HF, VHF, and UHF frequencies. Application experience includes access control, baggage/package/pallet tracking, livestock tracking, and WIP material tracking with installations in office buildings, warehouses, manufacturing facilities, delivery vehicles, and loading docks. A3 also has experience with EPC Class 0 and Class 1 for Wal-Mart and Target supply chain vendors. In the active tag arena, A3 specializes in RTLS (Real-Time Location Systems) and Asset Tracking applications, providing precise location capability and movement history for high-value objects on production lines, in hospitals or even parking lots.

Aanza AutoID Group

Aanza AutoID Group is a business and technology consulting firm enabling companies to optimize RFID usage and maximize the ROI while meeting Department of Defense, Wal-Mart, and other mandates. Services include business case development, process optimization, vendor selection, project planning, and implementation.

Accelitec

Accelitec provides RFID transponders and readers, self-service issuance, and management of RFID key tags for retail customers. The Bellingham, Washington, company also offers software for loyalty programs, point-of-sale integration, back-office support, and customer service and monitoring.

Accusort Systems, Inc.

Accu-Sort has specializes in automatic identification technology, including barcode scanners, CCD vision systems, and RFID technologies. The company, based in Hatfield, Pennsylvania, has helped its clients streamline their operations by managing materials and collecting and analyzing data. Accu-Sort's FAST Tag RFID system incorporates barcode scanning, RFID labeling, RFID tag reading/writing, controls, and data management including communications with the client's WMS or ERP system.

AceIC Designs, Inc.

AceIC Designs is a system design and automation company providing solutions ranging from Application Specific Integrated Circuits to Applications for Supply Chain and Point-of-Sale. AceIC offers off-shore execution modeling coupled with its R&D expertise.

Acheson Colloids Company

Acheson Colloids Company, based in Port Huron, Michigan, manufactures electrically conductive PTF inks and SMT adhesives. It is also an RFID label printer supplier and provides technology consulting.

AeroScout

AeroScout provides enterprise visibility solutions that bridge the gap between Wi-Fi, RFID, and GPS. AeroScout enables standards-based location and presence-based applications for indoor and outdoor environments where real-time visibility of assets and people is required to drive revenues or cut costs. The company's Wi-Fi-enabled tags, location receivers, and software use TDOA location to accurately track assets (including any wi-fi device) within large networks, without huge infrastructure investments.

AgInfoLink

This Longmont, Colorado, company provides agricultural information across the supply chain by utilizing RFID and other technologies, and integrating with existing MES and ERP systems, to provide traceability solutions for improvement in efficiencies and profitability.

Alanco Technologies, Inc.

Alanco headquartered in Scottsdale, Arizona, is the developer of the TSI PRISM RFID continuous tracking system for the corrections industry, which tracks the location and movement of inmates and officers, resulting in significant prison operating cost reductions and enhanced officer safety and facility security. Utilizing RFID tracking technology with proprietary software and patented hardware components, TSI PRISM provides real-time inmate and officer identification, location, and tracking capabilities both indoors and outside. TSI PRISM is currently utilized in prisons in Michigan, Illinois, and California, and in Ohio.

Alien Technology

This company is a supplier of RFID hardware that enables consumer packaged goods companies, retailers, and other industries to improve their operating efficiency throughout their supply chain. Using a manufacturing process called Fluidic Self Assembly, Alien fabricates EPC Class 1 tags in high volume at low cost. Alien's RFID readers provide a wide range of options, from industrial readers for the supply chain to compact reader engines suited for handhelds and printers. These products are interoperable with products from other Class 1 vendors, ensuring ease of implementation and multiple sources of supply. Alien is headquartered in Morgan Hill, California.

Analytica India

Analytica India provides an RFID-based asset tracking and location management system for hospitals. The system features customizable events, a floor layout viewer, search and filter, and asset movement history and reporting. The Bangalore-based company also offers RFID technology consulting, evaluation of RFID products, business process analysis for pilot programs, custom application design and development, integration services for advanced pilot deployments, and ROI evaluation for various domains.

Argent Group

Headquartered in Troy, Michigan, Argent provides label conversion, systems integration, and middleware across the complete RFID frequency range. Argent focuses on manufacturing operations in the pharmaceutical, automotive, food and consumer packaged goods industries, but works in the asset/documentation tracking and security areas as well. Argent also offers capacity for the insertion of RFID inlays into label stock for trade partners. The company can further assist clients with RFID business solutions that utilize on-demand thermal printer/encoders, or pre-printed pressure-sensitive custom labels for large and small users.

Assyst Inc.

This McLean, Virginia–based company has been in operation for 1 years, providing application software support, ERP implementation and integration support for a variety of commercial and government clients in the United States, Europe, Middle East, and Asia. Assyst helps customers comply with customer mandates as well as put in place an effective RFID setup.

AVANTE International Technology, Inc.

This Princeton, New Jersey, company offers interconnection technology for heat- and moisture-resistant RFID tags using HF, UHF, and other frequencies. AVANTE also offers 13.56 MHz antenna arrays that verify authenticity on RFID tags with a relational check-code. The company's Leads-Trakker trade show and event management system provides automatic attendance audit and lead retrieval.

Avicon

Avicon is an RFID systems integrator with experience in the use of RFID for open supply chains. It provides architecture development, technology vendor selection, systems integration, and project management for RFID projects.

AXCESS Inc.

AXCESS provides RFID systems for physical security and supply chain efficiencies. The battery-powered (active) tags locate, track, monitor, count, and protect people, assets, inventory, and vehicles. Also, the company has increasingly been called upon to leverage its RFID technology in wireless "sensing / monitoring" applications such as temperature monitoring, pressure monitoring, radiation/haz-mat detection, and open/closed condition monitoring. AXCESS active RFID solutions are supported by its integrated network-based, streaming digital video (or IPTV) technology. Both patented technologies enable applications, including automatic "hands-free" personnel access control, automatic vehicle access control, automatic electronic asset management, condition monitoring/sensing, and network-based security surveillance. AXCESS is a partner of Amphion Capital Partners LLC.

Baxter Healthcare

Based in Round Lake, Illinois, Baxter offers the QuickFind Asset Management System, which utilizes RFID technology to determine the location of any tagged asset. The system consists of a dedicated communication network installed above ceiling height throughout a facility, connecting transceivers that provide required coverage. The system software computes concurrent asset locations, which can be displayed on one or multiple PC's. The system's ability to track and locate assets quickly provides improved staff productivity, improved maintenance and calibration management, enhanced asset utilization, lower inventory costs, and reduced asset shrinkage.

Bibliotheca RFID Library Systems AG

This Swiss company targets libraries with its RFID system, which includes automated book return, patron self-check units, access control, staff stations, automated inventory, smart cards, and ISO 15693 labels. The system is non-proprietary and features batch-processing capabilities.

Bielomatik Leuze GmbH + Co. KG

This German company manufactures RFID transponder processing systems. Bielomatik's smart labels, tags, and tickets feature low per-unit cost, high quality, and expandable capacity. Entry-level solutions are available that can be upgraded to higher volumes at a later stage.

BT Syntegra

The company's Auto-ID Managed Service uses data from RFID readers, bar codes, and other signaling technologies to help manufacturers and retailers increase speed-to-market, avoid out-of-stocks, and lower operating costs. BT Syntegra's system includes technology selection support; a Web interface for monitoring, capturing and integrating data with existing supply chain applications; and implementation and support at either the client site or a BT Syntegra data center. The Arden Hills, Minnesota, company also helps clients evaluate ROI and comply with requirements from global retailers such as Wal-Mart and Tesco.

CapTech Ventures

This company uses software engineering processes to solve complex business problems for a wide range of clients. CapTech has released TagsWare, software that eases the deployment of RFID-enabled applications. The Richmond, Virginia, company designed TagsWare to speed compliance with Wal-Mart and Department of Defense initiatives; provide tag, reader, and platform independence; and help companies keep up with changing RFID standards.

Catalyst International

Catalyst International delivers software and solutions that enable companies to optimize the performance of their enterprise supply chains. The 25-year-old company offers warehouse and logistics software development, has an in-depth understanding of ERP systems, and assists customers with planning, installing and deploying integrated RFID solutions. Catalyst customers include Boeing, Brown Forman, Office Max, Panasonic, Rayovac, Reebok, Subaru, and The Home Depot. It is headquartered in Milwaukee, Wisconsin, and has offices or representatives in the United Kingdom, Mexico, and South America.

CCL Label

Based in Upland, California, CCL Label has provided pressure-sensitive adhesive labels and promotional products for the packaging, promotional and pharmaceutical industries for more than 50 years. The company also designs and prints RFID and anti-counterfeiting labels. CCL Label has plants located throughout North America, Europe, and Asia.

Chariot Solutions

Headquartered in Fort Washington, Pennsylvania, Chariot specializes in integrating RFID readers and software with existing Java application servers and enterprise software systems. The company provides strategic direction in choosing RFID hardware and software, helping clients develop an ePC-compliant RFID pilot implementation. Chariot also assists manufacturers and distributors with business process improvements that will enable them to capitalize on the benefits of RFID technology.

Checkpoint Systems, Inc.

Thorofare, New Jersey–based Checkpoint Systems is a multinational manufacturer and marketer of technology systems for retail security, labeling, and merchandising. Checkpoint is a leading provider of radio frequency-based electronic article surveillance systems. In addition to its EAS systems for shrink management, Checkpoint offers supply chain RFID technology to help apparel and consumer product manufacturers and retailers brand, track, and secure goods worldwide. Checkpoint has a presence in more than 50 countries and a global network of 29 service bureaus located in the world's apparel manufacturing capitals. A member of EPCglobal, Checkpoint's products include digital RF/EAS and EPC/RFID systems, RF source tagging, barcode labeling systems, EAS, handheld labeling systems, and retail merchandising systems.

Chemsultants

Chemsultants provides development and pilot services for conductive inks and polymer systems used in antennae printing for a variety of RFID tag and label applications.

China Elite Technology Company Limited

CET develops and manufactures radio frequency identification (RFID) products. CET is a subsidiary of Group Sense Ltd. (GSL), an electronics manufacturer publicly listed in Hong Kong. CET offers RFID products ranging from contactless smart cards and readers for access control and employee identification applications, to long range active and passive UHF RFID tags and readers for logistics tracking and inventory management. In conjunction with GSL, CET provides OEM and ODM services to major global RF clients. CET possesses industry expertise and China operational experience, combined with ISO9000 and ISO14400 certified production facilities.

CMS Consultants

CMS Consultants is a logistics management innovator, providing transportation, shipping and logistics solutions to companies around the world. CMS Consultants' focus is on the information management aspects its customers' shipping solutions. WorldLink, CMS's flagship enterprise middleware shipping solution, supports multi-carrier shipping and offers rating engine independence. WorldLink solutions centralize and simplify systems administration, enable global visibility of shipping activity, provide tools for companies to analyze their shipping data, and offer an array of productivity and customer service enhancements. WorldLink's RFID module allows companies to generate shipment-specific RFID tags.

CoBaLt Technology

Headquartered in Savannah, Georgia, CoBaLt specializes in condition-based logistics (CBL). The company offers active RFID tags with integrated environmental sensors; combined with its tag-to-tag networking scheme, the result is a full in-the-box-visibility system that can be easily integrated with global tracking systems. CoBaLt has also developed a passive-to-active RFID device to provide full visibility of products.

Cognizant Technology Solutions

Cognizant is a systems integrator offering an RFID solution that includes applicability analysis and scenario modeling to identify the areas for RFID implementation: selection of pilots, technology and vendor evaluation, cost-benefit analysis, and implementation of pilot projects and full-scale implementation. Its middleware hardware vendor and data format is independent. It has a layered architecture with separate layers for data collection, filtering, translation, and integration with other applications. It also provides device management functionality to configure, monitor, and control arrays of readers and other middleware components.

Comtrol

This Minneapolis-based company focuses on serial device connectivity, access, and control. It offers a range of hardware and software products and supporting services, from in-server multiport cards to network-attached programmable device servers and application appliances.

ConnecTerra

Enterprise infrastructure software is the focus of this Cambridge, Massachusetts, company. ConnecTerra's technologies provide communications, security, policy, device, and data management services required to integrate devices into a wide range of applications. The company's flagship product, RFTagAware, targets the challenges of enterprise RFID tag deployment, including tag data processing and reader management.

CopperEye, Ltd.

CopperEye provides software that allows very large flat-files of RFID transaction information to be accessed as a high-performance data repository. CopperEye complements relational database architectures by moving large volumes of transaction information to lower-cost file system storage without sacrificing any accessibility.

Cougaar Software, Inc.

This McLean, Virginia–based company features ActiveEdge software. Cougaar's product collects data from sensor hardware such as RFID, bar code, button memory, and physical devices, and integrates it into enterprise applications such as ERP, SCM, and transportation and warehouse management. The technology allows for easier data acquisition and management automation, better fine-grained planning and improved real-time information analysis. ActiveEdge enables enterprises to create an operational understanding of events such as RFID signals as applied to a business context.

Covansys

Based in Farmington Hills, Michigan, Covansys offers system integration and remediation for international and domestic asset tracking issues. Solutions include regulatory compliance under the Homeland Security, Patriot, and Sarbanes-Oxley acts through the entire supply chain for assets and financial transactions subject to those mandates. The company's services include business process evaluation, workflow design, portal management and automated exception management to ensure timely and complete delivery of data and goods within regulatory guidelines.

Craig Lamb & Singletary, Inc.

Based in Birmingham, Alabama, Craig Lamb & Singletary specializes in vendor-neutral RFID solutions for financial-services companies, including loan process tracking and lockbox management systems. It also provides business process management and business activity monitoring using business intelligence and Six Sigma methodologies. The company's technology analyzes and processes historical data to identify trends, uncover opportunities to increase revenue and reengineer processes across the enterprise, and turn passive applications into self-managing, intelligent systems that allow users to focus on the outcome of critical business processes.

Data Technology Group, Inc.

DTG features RFID products in the ePCGlobal (UHF 915 MHz) and dual-frequency (126 kHz to 7 MHz) sectors. The DF tag system allows reading through concrete and other solid materials, liquids, and body matter. The Atlanta-based company also offers GPS and cellular-based tracking products.

DC Logistics

This Hutchins, Texas, company operates the RFID Deployment Center in conjunction with SIS Technologies and Transport Industries. The center enables Wal-Mart suppliers to comply with regulations for shipping into the retailer's RFID Pilot Distribution Centers in north Texas. These services will expand to facilitate compliance with future RFID mandates from Wal-Mart, the Department of Defense, and others. Additional capabilities will include tagging, cross-docking, and consolidation services.

DDK International Inc.

DDK International is a systems integrator based in Richardson, Texas. DDK International provides customized integration and support services. The company uses and supports open source software and EPCGlobal standards so that your RFID system will be compatible in the industry.

Defense Systems Inc.

This Manassas, Virginia–based company provides RFID solutions that comply with the Wal-Mart, Department of Defense, and other mandates. Defense Systems features strategic consulting, product testing, and site analysis; it is also hardware- and software-agnostic, offering its clients best-of-breed products for the required implementation. The company also offers non-supply-chain RFID solutions such as asset and inventory tracking.

Distribution Management Systems Inc.

Distribution Management Systems (DMS) provides computer solutions (ERP, warehouse, logistics, purchasing, AP/GL, RFID, bar code, RF systems, and voice) to processors and distributors of food (including produce, meat, tobacco, candy, and beverages) and food equipment, and to wine and spirits distributors and importers. Since its inception in 1979, DMS has been a leader in the design and implementation of computer systems for distributors.

DYNASYS

Based in Clearwater, Florida, DYNASYS is the U.S. distributor and support center for RFID products from Texas Instruments, IDmicro, Precision Dynamics, Zebra, Impro Technologies, and other vendors. DYNASYS also manufactures its own line of RFID products to supplement the items from its vendor roster.

ecVision

This Newark, New Jersey, company provides integrated RFID packages that include software, hardware, tags, and implementation services to help clients in the retail sector gain maximum benefits throughout the supply chain. The system supports RFID tagging at carton/case and pallet levels, and can be implemented at the source of production.

Ekahau

Ekahau makes a real-time locating system, The Ekahau Positioning Engine and T-101 WiFI Tags (or Ekahau Client 3.x software tags) that use the industry standard 802.11 networks to track the location of assets and people.

Emerson & Cuming

Emerson & Cuming manufactures adhesives and encapsulants for circuit assembly and protection. The company's snap cure conductive and nonconductive adhesives are used to bond both bare dies and die straps onto antennas to manufacture low-cost RFID tags and smart labels, and its snap cure encapsulants are used to provide added protection to the RF circuit. Emerson & Cuming can optimize formulations to meet a company's unique process and performance requirements.

Encore Graphics

Based in Beaverton, Oregon, Encore Graphics is a distributor of RFID labels for inventory control and supply chain tracking.

Enterprise Information Systems

This Dallas-based company offers RFID/EPC integration services that help accelerate supply chain visibility, provide policy compliance, and improve overall efficiency. These services include readiness assessments, on-site RFID seminars, hardware analysis, and full system integration. Typical applications improve operations in the warehouse, on the manufacturing floor and in the field for mobile workers. EIS features its Get Ready, Get Set, GO! RFID implementation methodology.

EPC Integrator

EPC Integrator, headquartered in New Haven, Connecticut, is an RFID solution provider that offers a suite of products called EPCFusion that is aimed at creating business value for customers through strategic fusion of EPC data with enterprise information (ei). Its constituent products implement the various components in the EPCglobal Network architecture and extend it to create business applications. EPC Integrator's (ei) Middleware server and (ei) Information server provide the technology foundation, while the (ei) Business server implements various business functions such as asset tracking and genealogy. The company also offers vertical solutions for retail and healthcare built on top of their Business server.

EPCglobal

EPCglobal is leading the development of industry-driven standards for the Electronic Product Code™ (EPC) to support the use of Radio Frequency Identification (RFID) in today's fast-moving, information rich, trading networks.

They are a subscriber-driven organisation comprised of industry leaders and organisations focused on creating global standards for the EPCglobal Network™.

Their goal is increased visibility and efficiency throughout the supply chain and higher quality information flow between companies and their key trading partners.

epcSolutions, Inc.

Great Falls, Virginia–based epcSolutions offers an EPC/RFID middleware product called ThingsNetTM, which aims to help enterprises reduce the time, cost, and effort of RFID projects and fully meet Wal-Mart and DoD compliance. ThingsNet consists of an EPC-compliant Savant, which handles read events and interfaces with external applications, and an EPC Information Service, which stores tag data and containment relationships between tags (i.e., cases on pallets) while linking to both the EPC/RFID network and back-end applications. The ThingsNet Control Center is an EPC/RFID network management application for monitoring and controlling Savants and readers.

Escort Memory Systems

Escort Memory Systems (a Datalogic Group company) provides RFID solutions for every link of the supply chain. Since 1985, EMS has been developing, manufacturing, and supporting RFID installations in companies around the globe. Supply chain customers with applications such as work in progress, quality control, warehousing, and logistics use EMS RFID for inventory management, tracking and data collection systems. EMS uses technology designed to read through water, oil, concrete and a variety of other elements without line-of-sight requirements, or waiting for each individual tag to be read.

eXI Systems

eXI develops, manufactures, and markets RFID products and services, including Assetrac. Using one of the smallest and longest-lasting active tags available today, Assetrac provides asset protection and real-time location. Other products include RoamAlert, a system that prevents Alzheimer's and dementia patients from wandering away from nursing homes and assisted-living facilities, as well as the HALO Infant Protection System, a system that helps prevent infant abductions in hospitals. eXI, through its subsidiary HOUND-ware, has implemented tool-tracking systems in such industries and sectors as construction, building management, equipment rental and management, natural resources, power and utilities, and chemical processing.

Fast Forward Technologies

Based in Louisville, Kentucky, Fast Forward Technologies manufactures RFID hardware and software tools for application developers and provides custom application development services. The Fast Forward product line includes infrastructure components including readers, middleware, and a complete application framework.

FileTrail, Inc.

FileTrail is a solutions provider for records management that automates tracking to provide timely location and retrieval of records using bar code and RFID.

FKI Logistex Baggage Handling Team

The FKI Logistex Baggage Handling Team provides RFID and explosion-detection systems integration for small and regional airports. As its name suggests, the Louisville, Kentucky, organization is also a leader in integrated baggage-handling solutions worldwide.

Franwell, Inc.

This Florida company is a leader in the research, development, and implementation of RFID technology for supply chain operations. The company's RFID Genesis system, combined with its Agware enterprise software, enables clients to be EPC-compliant while retaining the ability to adapt to changing standards. Franwell is engaged in a project with the University of Florida's IFAS Research Center for Food Distribution and Retailing to test the use of RFID technology as it relates to each link of the food supply chain.

General Data Company, Inc.

The company provides RFID and smart-label equipment and systems, including decoders, labels, tags and printer/encoders. General Data, headquartered in Cincinnati, partners with industry-leading manufacturers including Zebra

Technologies, Texas Instruments, and Symbol Technologies to bring products to complete RFID printing solutions.

GenuOne

GenuOne's TraceGuard middleware is an item-level product tracking platform for RFID systems and other tagging technologies, serving such industries as consumers' packaged goods and pharmaceuticals, as well as DoD suppliers and well-known consumer brands. GenuOne also provides professional services to develop RFID business case and conduct RFID assessments, pilots and rollouts.

GlobeRanger Corporation

Based in Richardson, Texas, GlobeRanger provides iMotion, a software platform that simplifies the development, deployment, and management of integrated RFID/barcode, mobile and sensor-based systems. The platform gives developers and integrators the ability to quickly create solutions by providing a library of applications components and device adapters, visual workflow management, and standards-based integration. iMotion provides a software infrastructure layer that networks data-collection devices and back-end systems to manage the continual flow of information, enabling alerts, decision support, and real-time response.

GrowSafe Systems Ltd.

GrowSafe Systems has been developing RFID solutions since 1990 for the agricultural research, cattle, and dairy industries. The company is both a systems integrator and OEM with significant engineering expertise. GrowSafe systems are installed in major agricultural institutions and high-capacity livestock organizations. GrowSafe has patented a method to read multiple passive transponders in close proximity. GrowSafe is one of the few RFID technologies in any industry that use multiple antenna arrays positioned to cover all the possibilities of reading a tag. It has designed its systems to automatically ensure that an animal is positioned in an optimal reading range without human intervention. A GrowSafe system can collect data from 65,000 nodes simultaneously and wirelessly transmit this data over a 30-mile radius to a central computer. The company has also developed intelligent software agents that identify sick animals, nonperforming animals, and animals ready for market.

HAL Systems, Inc.

This Atlanta-based company offers its Tracker RFID-enabled warehouse management system for production and distribution environments. The system, which has both RFID and bar-code capabilities, uses RFID equipment from Texas Instruments, Hand Held Products, and LXE. Emphasis is placed on understanding the client's material flow and then matching real-time data collection requirements. Tracker is the cornerstone of HAL Systems' wholesale/retail Distribution-Express package, which includes integrated accounting, order processing, credit-card authorization, and shipping applications. Distribution-Express can be purchased as a complete solution or as individual building blocks that can be integrated with legacy applications.

Hitachi Europe Ltd.

Hitachi-Europe's RFID solutions are based on the world's smallest microchip, the Mu-chip, which measures 0.4-mm square. The U.K.-based company provides enterprise-wide RFID deployments, including readers, tags and software systems.

HK Systems

With more than 20-plus installations, HK has developed pioneering applications in the automotive industry as well as cross-sector integrated logistics management software. The Wisconsin-based company's goal is to bring practical, value-producing RFID solutions to its customers.

Horizon Services Group, LLC

Horizon Services Group delivers cost-effective "order-to-cash" logistics management technology through the integration of people, technology, and process. It assists clients to creatively use information technology to improve customer service while reducing the costs.

iAnywhere / XcelleNet

This Alpharetta, Georgia–based company's RFID framework is designed for enterprises and systems integrators that need to deploy and manage thousands of RFID readers, sensors and actuators and correlate millions of data points. Unlike RFID middleware that simply filters and routes tag data, iAnywhere/XcelleNet's product transforms the data into actionable information for concurrent use by multiple applications. It also provides real-time notifications to site, field, and corporate personnel.

IBM

RFID solutions from IBM allow companies to transform and optimize their supply chains, improve asset information management, and deliver increased levels of customer service. These programs include business case and deployment strategies, pilot planning and implementation, process design, tag/reader testing, and network and infrastructure design, deployment, and maintenance. IBM also offers open, scalable software products, as well as application and process integration systems.

IconNicholson

IconNicholson helps companies gain competitive advantage and long-term value by developing high performance solutions that maximize the power of digital and RFID technologies to build stronger businesses and customer relationships.

Identec Solutions, Inc.

Based in Kelowna, British Columbia, Identec Solutions is a global manufacturer of active tags, readers, and middleware. The company's RFID system, Intelligent Long Range, operates at 868 or 915 MHz; it identifies and tracks assets at distances of up to 300 feet. The system is designed for dynamic, demanding environments and is predominantly used for vehicle tracking and container management. Customers include General Electric, Volkswagen, Deutsche Post, and Thomas Built Buses.

IDmicro

This company provides industrial-scale, high-value asset and people-tracking systems to Fortune 500 companies. Its Remote Intelligent Communication technology enables reliable tracking with read/write ranges of up to 90 feet. IDmicro tags have a battery-powered chips and use backscatter technology, which enables the tags to operate at longer distances than passive tags. The company provides both hardware and software systems.

In2Connect Lt

In2Connect is an RFID antenna manufacturer based in Nottingham, U.K. The company has 35 years of experience in manufacturing flexible circuits for the RFID field. The company works in reel-to-reel processing of thin flexible substrates that can used for RFID inlets and antennas. In2Connect offers high volume processing on a variety of substrates including PET (polyethylene terephalate), Pi (polyimide), or PEN (polyethylene napthalate) together with thin copper conductors. Antennas can be supplied in reel format, either single or multi-antenna width, to suit customer's processing needs. Antennas can be supplied with or without dielectric coverlay films, or inks, and silver-printed conductor bridges. In2Connect's antennas can be used for a wide range of applications, including asset tracking and inventory control.

Infocom Systems

Infocom Systems provides custom solutions integrating technologies into next generation solutions. These technologies include speech, RFID, imaging, and more in order to help companies streamline operations, reduce costs, and improve customer service.

Innovative Equipment Ltd.

Based in Cortlandt Manor, New York, Innovative is the North American distributor of the Melzer SL line of automatic RFID label, tag, and ticket manufacturing systems. The Melzer SL series uses functional RFID inlays for application into labels, tags, and tickets. Each machine has both HF and UHF capability and can run virtually any inlay on the market at production speeds of up to 32,000 units per hour.

Innovision Corp.

Innovision offers NEPAL Developer, a tool used for protocol engineering: the process of validating and creating new or enhancing existing XML protocols to standardize and simplify information exchange. NEPAL Developer is a subset of the NEPAL protocol technology platform, a framework for building and executing protocol-based systems. With NEPAL Developer, high-level protocol development and deployment becomes a process of assembling reusable components to efficiently and securely deliver information to a variety of RFID-enabled endpoints.

Innovision Research & Technology P.L.C.

Based in Wokingham, U.K., Innovision designs, develops, and licenses RFID solutions. Its custom IC design group has produced more than 50 RFID tags and readers. Innovision supports the entire development process from concept to production engineering.

INSIDE Contactless

This semiconductor company focuses on the development of contactless chips and readers for emerging markets such as transportation, access control, payment, identity, and electronic identification. Inside Contactless provides clients with a complete contactless technology platform. Headquartered in Aix-en-Provence, France, with offices in Shanghai, China, and Wilmington, Delaware, the company works with systems integrators, VARs, and application developers.

Institute of Applied Physics, Department for Automation

Based in Russia, the Institute offers an active-tag RFID system with an effective control area of 1 square kilometer (outdoors) or 1,000 square meters (indoors). Tags can be read at distances up to 500 meters, with a positioning accuracy of 1 meter. Maximum number of tags is 10,000.

Integral RFID Inc.

Integral RFID specializes in Wal-Mart- and DoD-mandated RFID solutions. It works with the leading software integration teams to efficiently cover the entire technology spectrum from RFID equipment recommendation and acquisition, to installation and full operational integration with existing warehouse management system (WMS).

Intellareturn Corp.

This New York–based company provides logistics solutions to identify and return virtually any object. Intellareturn's wireless shipping and notification systems create real-time interaction between products and parcel shipping networks, using special labels to automate identification and trigger the return of any item. This interaction, via the Internet or a LAN, helps reduce expense associated with wasted or delayed product returns, as well as lost or stolen items. The company's systems use a distributed services model that can link to virtually any legacy parcel or postal service network.

Intermec Technologies Corp.

RFID products from this Everett, Washington–based company are used in a wide range of government and commercial applications. These include manufacturing, food processing, security, and logistics in the United States; retail and logistics in Europe; biohazard waste management in Africa; and transportation systems in Japan. Intermec's UHF readers and tag systems are certified for operation in the United States at 915 MHz and 2.45 GHz, in Europe at 869 MHz (meeting narrowband requirements), and in Japan at 2.45 GHz.

International Paper

International Paper is the world's largest paper and forest products company. Its businesses include paper, packaging, and forest products. Headquartered in the United States, International Paper has operations in more than 40 countries and sells its products and services in more than 120 nations. Its Smart Packaging unit offers EPC-enabled supply chain solutions from warehousing and transportation tracking to the retail shelf. The company designs, developments, and implements integrated RFID packaging and supply chain applications with an emphasis in the Electronic Product Code (EPC).

iPico Identification

The iPico Identification UHF, 2.45 GHz, and dual-frequency RFID products offer long-read-range, fast multi-read, anti-collision, and high through-beam detection of dynamic tag populations. Ipico produces smart labels and tags based on the company's iP-X communications protocol. The low-interference readers comply with regulatory requirements in major geographical regions, and allow for multi-reader rollout in close proximity. The iPico DIMI middleware platform manages applications in large-scale, distributed systems; the DIMI data and connectivity architecture allows for effective device management, data switching and data security.

Irista (HK Systems-Irista)

HK Systems-Irista is an automated material handling equipment and supply chain software provider. It offers applications in the automotive industry and cross-industry integrated logistics management software. The company has 17 years of RFID experience and has done more than 20 RFID installations.

Ito America Corp.

Ito America Corp. (IAC) is based in Scottsdale, Arizona, and offers the following products and services: Sony Chemicals anisotropic conductive films (ACF), Nippon graphite heat-seal connectors, polarizer film (as well as removal and laminating equipment), Ohashi ACF processing equipment for displays and display flip-chip applications, Shibaura flip-chip bonders for semiconductor ACF flip-chip applications, Nissha polyimide printers, rubbing lines and align/mate tools, interconnect consultancy services, interconnect prototype and testing services, and offshore high-volume manufacturing for ACF applications.

KartKeeper

KartKeeper is a system designed to make it easy to manage returnable container investments. Using radio frequency technology, KartKeeper employs a hands-free capability to tracking containers to and from customers. Employing intuitive software screens to allow for easy access to company shipping information, the KartKeeper system tracks carts from the shipping dock to the customer and back again.

Laudis Systems

Laudis Systems offers turnkey RFID-enabled tracing, tracking, and monitoring solutions. Applications include material and asset tracking, perishable goods monitoring, and transportation and vehicle management. The system is

designed to support both passive and active tags, including those with integrated sensors.

Leader Induction Technology

Leader Induction Technology produces designs and manufactures smart cards,
IC cards, RFID tags, and labels. Its products are used in automated parking
payments, loyalty cards, banking, access security, data management, personnel,
and animal/bird identification. It supplies many European conglomerates.

Lorantec Systems, Inc.

Lorantec Systems provides real-time location tracking and monitoring of
cargo and equipment assets worldwide. It was founded in 2002 and headquartered in Sunnyvale, California. The company's flagship LoranTrack service
combines intelligent asset tracking and monitoring devices, Low Earth Orbit
and GPS satellite constellations, and web-based reporting and event notification software.

Lowry Computer Products, Inc

Lowry Computer Products, a nationwide provider of RFID-EPC, bar code,
wireless, and data collection solutions for the Supply Chain, offers EPC-compliant case and pallet solutions, digital and EPC-enabled "smart" labels, automated labeling systems, EPC readers, EPC encoding and printing systems,
complete installation services, and 24 × 7 technical support. Lowry's application expertise includes compliance labeling, product and case labeling, shipping and receiving, warehouse and inventory management, and more.

Lyngsoe Systems

Lyngsoe Systems is a systems integrator and software manufacturer with
offices in Denmark, Canada, and the United States. It offers a wide range of
IT solutions for monitoring and automating processes in complex logistics
chains. During the last 10 years, Lyngsoe Systems has developed and installed
a global RFID network for a postal supply chain covering more than 50 countries and more than 600 distribution centers. The company is independent of
specific technologies and vendors, which enables Lyngsoe to select the technologies to match the specific requirements of customers.

MagTech Systems, Inc.

This Carmel, Indiana–based company offers RFID process-improvement
and software solutions. MagTech Systems is a certified Intermec and Alien
partner.

Manhattan Associates, Inc.

This Atlanta-based company's RFID in a Box product offers RFID technology, supply chain execution applications and professional services to simplify implementations and enable vendors to meet compliance requirements. It includes RFID readers with two antennas per reader, as well as tags from Alien Technology to label and track goods. The package also offers a limited license version of the Trading Partner Management application, which allows customers to enable their suppliers to remotely generate RFID tags and apply them to goods.

Mannings RFID

Headquartered in Merseyside, U.K., Mannings RFID is a systems integrator that offers solutions for asset tracking, access control, baggage handling, and other applications. The company also customizes existing products to meet its customers' needs.

Markem Corp.

The MARKEM mission is to help industry mark, code, or identify products through innovative marking equipment, software, supplies, and service.

Matrics Inc.

Matrics designs and manufactures EPC-compliant RFID systems. With its partnership network, Matrics provides RFID solutions to many Fortune 1000 companies and to government agencies. Current customers include International Paper and McCarran Airport.

Microsoft Corporation

Microsoft plans release in 2007 of their RFID Services Platform, a "middleware" product that connects the hardware that monitors RFID signals with the business software that can make sense of the information. The product is designed for businesses that want to incorporate RFID into their own systems, as well as for other software companies that want to build a product based on Microsoft's technology.

The RFID product will be built on top of Microsoft's .NET development platform and will run on a two-processor server. It will also incorporate the company's SQL Server database software for information storage.

Miles Technologies

A systems integrator headquartered in the Chicago area, this company has more than 20 years' experience providing government and commercial clients with inventory control systems, integrated data collection devices, RFID, wireless systems, barcode printing, custom programming, and software. Miles is certified by Alien Technology, Intermec, and Zebra Technologies.

MPI Label Systems

This Sebring, Ohio, company manufactures RFID labels and tags, either pre-printed or blank, on paper and synthetic materials. The tags come with transponders that work at 13.56 MHz, UHF, and other frequencies. MPI offers a 360RW RFID label applicator that inspects transponders, writes real-time information during application, and rejects malfunctioning transponders. The company also sells tabletop RFID label printers and automatic RFID printer-applicators, all with RFID read/write features and the ability to reject bad transponders.

N.C. Cuthbert

N.C. Cuthbert is a New Jersey–based systems integrator that offers chassis, container, and genset tracking for ocean carriers and terminal operators. The company's systems automate seaport and container yard operations, including appointments for truck visits, verification of mission, security control, and traffic queue control. It offers integration with existing systems, replacing key entry, and has experience with video-based OCR, voice synthesis and traffic information radio, RFID, laneside driver communication pedestals, and weigh-in-motion scales. Cuthbert clients' business partners include truckers, brokers, forwarders, and shippers.

National Service Center

National Service Center, based in Greenville, South Carolina, specializes in on-site service and installation of auto-id equipment including RFID, barcode, datacapture, and POS.

NCR Corporation

With 120 years of experience serving retailers, NCR is focused on facilitating the adoption of RFID in the retail environment. The company creates and tests proof-of-concept prototypes using RFID and other advanced technologies and works with RFID technology vendors on next generation RFID-enabled systems. Through pre-modeling work, Teradata, a division of NCR, can manage additional data that will be available as RFID is integrated into the retail industry. NCR Managed Services offers capabilities from initial site assessments through project management, installation and deployment, as well as a full suite of maintenance and help desk services. NCR's Systemedia Division has a core competency in printed consumables and uses thermal transfer printing technology to create RFID-equipped labels.

Northern Apex—RFID

Indiana-based Northern Apex is a technological partner with several RFID technology providers operating at multiple frequencies and is an integration partner with TagSys and TI for LF 13.56 MHz and with Matrics for UHF 900 MHz solutions. In addition to providing tags, reader boards, and antennas, Northern Apex's Product RFID data capture devices work with multiple technologies provided by TI, Phillips Icode, and TagSys 13.56 MHz tags, as well as the ISO15693 tags. The company is affiliated with both Symbol and Palm partner groups for work on handheld designs. Northern Apex has also developed software products with do it yourself library command sets and complete, specific application solutions.

ObjectStore

This Bedford, Massachusetts–based company's flagship product is the Real Time Event Engine, which features a real time in-memory database (REID) with EPC code support, and Savant RFID interfaces. The REID uses EPC-global architecture to provide consistent, scalable and fault-tolerant EPC event processing. ObjectStore's Event Engine also supports EPCglobal standards for tag formats, specifically the 96-bit Class 1 and Class 0 EPC codes defined by the Version 1.0 specification.

Octave Technology Inc.

This College Park, Maryland, company manufactures and distributes RFID-enabled asset maintenance software for use in rugged and industrial settings such as power plants, manufacturing facilities, data centers, repair depots, and fleet and storage centers. Octave's products and services help industrial asset owners effectively manage, maintain, and secure their infrastructure and equipment; ensure the safety of service engineers and technicians; and reduce operational inefficiencies inherent in manual processes.

Open Tag Systems

Based in Colorado Springs, Colorado, Open Tag Systems offers low-cost, ergonomic hand-held RFID readers. A wide range of reader interfaces allows the company's products to be integrated into existing systems. OTS products can be used in point-of-sale, manufacturing, warehousing, and other applications.

ORACLE

Oracle is the world's largest enterprise software company. For more information about Oracle, visit their Web Site at http://www.oracle.com/. Utilizing Radio Frequency Identification (RFID) technology, companies can more accurately track assets and monitor key indicators, gain greater visibility into their operations, and make decisions based on real-time information. Increasingly, RFID tags are being combined with many types of sensors and tracking technologies like GPS to give companies greater visibility into their supply chains for reduced risk and optimized business processes.

Based on Oracle Fusion Middleware, Oracle Database 10g, and Oracle Enterprise Manager 10g, Oracle Sensor-based Services enable companies to quickly and easily integrate sensor-Based information into their enterprise systems. Oracle's solution includes a number of flexible deployment options that allow companies to start small and grow by leveraging their investment in Oracle technology, including the Supplier Compliance Workspace an integrated support element of the Oracle E-Business Suite and Oracle Applications Server 10g. Oracle's solution includes a Compliance package, an RFID pilot kit, and integrated support in Oracle E-Business Suite and Oracle Application Server. Oracle is a member of EPC Global, which is leading the development of industry standards for the Electronic Product Code (EPC) Network to support the use of RFID.

Panther Industries Inc.

Panther Industries is a manufacturer of automated labeling equipment. Products include the Panther 2000 & Panther 2000e Print and Apply systems and the Panther CUB Label Applicator. These systems feature the standard Operator Interface display for operator feedback and control with 100% off-the-shelf components. There are no proprietary controllers, PC boards, or power supplies in our systems. The entire line can be purchased RFID enabled. Using print technologies from Datamax, Sato, and Zebra, the Panther 2000 Prints-Applies-Encodes 900 MHz EPC labels. Currently it uses the EPC Class 1 standard, but is upgradeable to new standards as they emerge. These systems can also be purchased with an automatic tag reject mechanism for situations that required 100% readability or can be fitted with auxiliary readers for exception diverters.

Paratek Microwave, Inc.

Columbia, Maryland–based Paratek offers 900 MHz and 2.4 GHz Smart Scanning Antennas that support RFID and WLAN systems.

Parelec Inc.

Parelec develops and manufactures conductive inks, marketed under the Parmod trademark. Parmod inks are highly conductive, approaching the conductivity of etched metal antennas, and are delivered as either pastes or toners for screen, gravure, or flexographic printing onto paper or polymeric substrates for the manufacture of cost-effective RFID transponder and reader antennas.

Paxar Americas Inc.

Paxar products are used the world over by leading apparel brands. The company's Monarch brand products are used by 90% of the top 100 U.S. retailers and their supply chain partners to identify, track, and price all varieties of consumer goods. Paxar is a member of EPCglobal.

PharmaSeq, Inc.

Based in Monmouth Junction, New Jersey, PharmaSeq offers microtransponder-based systems that feature light-powered electronic tags. The microtransponders come in two versions, $500 \times 500 \times 150$ microns and $250 \times 250 \times 100$ microns. The tag is a monolithic integrated circuit built on a silicon chip. When illuminated by light, each tag's photocell provides power for the electronic circuits in the tag. Microtransponder readers are also available. The microtransponders can be built into a variety of documents, including plastic ID cards. Primary uses include security and product and document authentication.

Phase IV Engineering, Inc.

Phase IV Engineering provides active and passive RFID and sensing hardware and solutions. Phase IV's products sense temperature, pressure, humidity, and shock. Phase IV offers the CargoWatch family of active UHF tags, handheld and fixed readers, and software development kits for building applications that integrate CargoWatch components. Its Barrier Communication System (BCS) allows wireless transmission through metal with non-intrusive, remote wireless tracking from outside the container, the ability to monitor and control internally positioned door security devices, fully automated, and remote near real-time monitoring with optional integrated wide-area communication systems (e.g., SATCOM). CargoWatch products are designed to operate in harsh environments, and have low-energy emission levels.

Pittsfield Weaving Company

Pittsfield Weaving Co. has been supplying woven and printed labels to the apparel industry for over 78 years. The company has developed and holds key patents to the integration of RFID and EAS tags into apparel labels. This integration, according to Pittsfield Weaving, has proven to be the most cost-effective solution for tagging individual items of apparel.

Plitek

This Des Plaines, Illinois, company integrates RFID circuitry into a variety of labels and tags. Plitek uses QS 9000 quality standards for customers who require a Production Part Approval Process. The company's integrated manufacturing facility can handle small as well as large RFID applications.

Power Paper

Power Paper offers the PowerID System, integrating thin, flexible RFID labels with the company's patented, low-cost, printable power source. According to the company, the PowerID System provides the highest reliability of any RFID label system in the industry, read-only and read/write ranges (8–15 meters) that are far superior to those of existing RFID systems, and cost-effectiveness, with nominal cost of integration. Compatible with EPC, the PowerID System consists of UHF-based battery-assisted backscatter labels, readers and system software that integrates into and complements existing enterprise software. Power Paper believes that PowerID guarantees greater accuracy, reliability and simultaneous reading of multiple labels in the toughest RF environments, and that it also solves the problem of reading labels affixed to challenging products that contain metals, liquids, foils, etc., set at different angles, and sandwiched between crates.

Precisia

Precisia product offerings include low-cost printed electronics, including RFID antennas produced with conductive inks. Its ability to produce high-volume printed RFID antennas reduces the overall cost of RFID tags when compared with traditional copper-antenna or screen printed antennas. Precisia's offerings include fully enabled RFID devices, essential materials for RFID, and other printed electronics applications, including smart/active packaging, printed electronics, lighting, and displays.

Precision Systems

The iLocate tag identification and positioning system is the flagship product of this Tel Aviv, Israel, company. It combines RFID technology and optical surveillance tracking to enable tag data collection to an accuracy of less than 1 foot. Precision Systems also offers customized products, including RFID-enabled zone positioning systems and optical location systems.

Printronix, Inc.

Based in Irvine, California, Printronix offers the Smart Label Pilot Printer, a UHF Class 1 RFID commercial printer, along with the Smart Label Developer's Kit. The package is designed to fast-track RFID pilots by creating a smart label developer's environment. Printronix's RFID Software Migration Tools allow users to quickly move from printing UPC or GTIN barcode labels to encoding smart labels.

Provia Software

Since 1988, Provia has provided supply chain products for companies such as Gillette, Total Logistic Control, TaylorMade-adidas Golf, Menlo Logistics,

Lanier Worldwide, and Graybar Electric. The Grand Rapids, Michigan, company offers RFID-enabled warehousing, transportation and yard management products that are integrated with Web-based visibility and event management tools. Provia also offers a bolt-on RFID compliance solution for companies with existing WMS or host systems.

PSC Inc

PSC is a designer and implementer of open hardware and software systems that allow clients to manage the transition from bar code to RFID. PSC helps clients view the coming changes as evolutionary in nature, and part of a predictable, step-by-step sequence of changes as RFID begins to play a larger role of coexistence with bar codes. PSC's priority is to support solution provider's need for best-in-class rugged Mobile Hybrid Readers. These Mobile Hybrid Readers withstand extreme conditions, applicable across multiple classes of trade. PSC structures alliance partnerships to iteratively develop and verify RFID solutions.

Psion Teklogix

Psion Teklogix is a global provider of solutions for mobile computing and wireless data collection. PTX has deployed solutions for both low-frequency and high-frequency applications and has announced UHF availability for most of its product line. The company's solutions mobilize enterprise information, allowing employees to gather, enter, and share data at the point of work.

Quest Information Systems

Quest Information Systems helps clients take full advantage of the cost savings and increased revenues that can be realized with the use of RFID enabled systems. RFID enabled systems provide traceability in manufacturing plants to reduce manufacturing costs, increase product quality, reduce material management costs, reduce warranty costs, reduce out-of-stock instances, reduce shrinkage, and hinder counterfeiting by helping companies observe, understand, and manage the movement of goods in real-time.

R4 Global Services

This San Francisco company provides integrated RFID, GPS, and RTLS solutions to Fortune 500 and middle-market corporations in the retail, manufacturing, healthcare, logistics, and transportation industries. R4 has partner relationships with RFID vendors such as Matrics, Intermec, Alien, Texas Instruments, and WhereNet. The company also has expertise in ERP and supply chain applications, integration, and infrastructure.

Radcliffe Inc.

Radcliffe is a supply chain software company, specializing in warehouse management systems for corporate and third-party logistics operations. Radcliffe's RFID applications work with its WMS or as separate modules that can be integrated with WMS products from other vendors. Radcliffe integrates and reseller RFID equipment and smart labels from a number of manufacturers. The company also offers an Internet-based EDI engine that is integrated with its RFID offerings, meeting Wal-Mart's requirements for ASN communication.

Radianse, Inc.

Radianse, Inc., Lawrence, Massachusetts, provides indoor positioning solutions (IPS) designed for healthcare. The IPS is designed to reduce asset shrinkage and labor costs, improved patient flow-times and overall workflow efficiency. A Radianse IPS combines long-range active-RFID tags with a patent-pending location algorithm for accurate, continuous location and association of people, places, and things. Information is shared using Web and interface standards such as XML and SMS. Radianse receivers directly connect to a hospital's existing LAN. Healthcare organizations use the Radianse solution for asset tracking, patient location, and workflow management applications.

Rapidwerks LLC

Based in Pleasanton, California, Rapidwerks is a precision micro-molding company. It provides plastic components that can be insert-molded, over-molded, or co-molded with materials such as metals and other plastics, for RFID and related applications.

RealTime Technologies Inc.

RealTime Technologies offers a suite of enterprise solutions for real-time data capture and e-business integration within the manufacturing sector. According to the company, its ADC (automated data capture) solution, called RT/CIM, works out of the box for BPCS, Movex, Friedman, PRMS, and EAM.

RedPrairie Corporation

A member of EPCglobal, RedPrairie offers RFID systems and services including bolt-on compliance and processing solutions, and an RFID-enabled version of its SCE suite. The Waukesha, Wisconsin, company helps clients meet Wal-Mart and Department of Defense requirements.

ReturnMe.com, LLC

This New York company sells Return ID Stamps, postage stamp-sized smart labels that can be read by RFID devices to generate information via the Internet about warranty returns and expiration dates. ReturnMe is targeting parcel and express courier services, airline baggage offices, and retail outlets equipped with a reader and Internet-connected PC.

RF Code, Inc.

Founded in 1997, RF Code designs and develops RFID products including tags, readers, antennas, and software to manage and track physical assets, information, and personnel through the supply chain. The Mesa, Arizona–based company also offers TAVIS Auto-ID data collection middleware, which works with data collection devices including active and passive RFID readers, EPC readers, barcode readers, and GPS receivers. TAVIS converts reader data into a standard format that can be linked to end-user applications to track, locate, and identify assets.

RF IDentics

RF IDentics LLC, based in Grand Rapids, Michigan, offers a family of low-cost, high-performance RFID tags and labels. RF IDentics tags are ISO and EPC compliant and are available in UHF, 13.56 and 125 kHz. The principals of RF IDentics used the their combined 35 years of knowledge of running the world's largest material handling business to develop RFID tags for many applications: airline baggage, supply chain, medical, postal, and parcel.

RFID DataCorp

RFID DataCorp is an information technology company specializing in RFID technology. The company provides implementation of end-to-end RFID solutions, including design, implementation of RFID data collection systems, wireless communication, integration with backend systems, and EPC networks. RFID DataCorp specializes in providing end-to-end RFID solutions for the healthcare industry and develops solutions based upon open standards set by EPCglobal using off-the-shelf hardware components. Healthcare Solutions products include RF Real-time patient tracking, RF Asset Identification and tracking, RF Patient identification, RF Patient lab sample tracking, and RF Patient access control.

RFID Deployment Center

The RFID Deployment Center is a working warehouse with real products moving in and out. The center offers outsourced support for deployments to retail and DoD suppliers that are seeking to respond to the RFID initiatives through outsourced services, thus learning at a state-of-the art facility and mitigating the risk of implementing in-house too early.

RFID Solutions, LLC

RFID Solutions offers technical services include prototyping, technology selection, demos, pilot support, characterization of coverage, future-proofing RFID investment, and fit analysis with long-term strategic plans. It has several years of experience with RFID, both HF/13.56 MHz and UHF/915 MHz technologies from new, start-up, and established vendors.

RFID Sources Corporation

This Taiwan company offers 125 kHz and 13.56 MHz proximity clamshell cards, contactless ISO cards, hybrid cards, dual-frequency cards, thermal rewrite cards, RFID tags, and smart CDs. It can also provide clients with customized products such as key fobs, disc tags, and smart labels.

RFID Systems Inc.

RFID Systems's goal is to enable its clients to negotiate the complexities of implementing emerging technology and ensure increased operational efficiencies and long-range success through sustainable return on investment. Its RFIDpak warehouse management system has been credited with reducing labor costs by 38%, reducing stock outage by 10%, and improving overall the performance of distribution operations, according to RFID Systems.

RFID, Inc.

A developer of customized RFID tags and readers as well as off-the-shelf products, RFID Inc. offers free consulting services. Founded in 1984, the Aurora, Colorado, company's product line includes 125 kHz process controls (WIP), OEM-level solutions, access controls, micro-readers, and handhelds; 13.56 MHz ISO 15693 and 14443 solutions due for release in May; 433 MHz active tags for vehicle, personnel, and asset tracking; and 915 MHz EPC version 1.2-compliant readers (for Wal-Mart and Department of Defense projects) due in June. The 915 MHz readers are also backward-compliant to class 0 (Matrics) and class 1 (Alien).

RFideaWorks Corp.

RFideaWorks is an Oracle alliance partner that works with major systems integrators. The company's RFID director, designed to be totally interoperable middleware with all readers and any applications. Using a patent-pending

RFID tag translation, it handles EPC, ISO, UID, proprietary, and bar code standards, and can read thousands of tags per minute, according to the company. RFIdb is a high-performance database for processing high-speed/ large volumes of RFID tag data; it uses patent-pending in-memory processing to run at CPU speeds. RFIdwh processes terabyte-level RFID tag info at CPU speeds for real-time business intelligence. All solutions are integrated with back-end enterprise systems focus.

RightTag Inc.

Dedicated to providing products and services that enable companies to use radio frequency identification in a variety of ways, RightTag manufactures 13.56 MHz RFID readers and engines. It prides itself in providing its customers with inexpensive RFID solutions, for purchase or license, that integrate well with other systems.

RSI ID Technologies

Based in Chula Vista, California, RSI provides and supports RFID systems and process optimization for clients that range from mid-sized companies to Fortune 500 organizations. RSI offers RFID labels, tags and readers, consulting, strategic planning, process automation engineering, system design and integration, product marking systems, data collection solutions, warehouse management systems, real-time location systems, smart shelves, portals, custom software development, middleware integration, support services, and training. RSI's customers are in sectors including manufacturing, logistics, distribution, warehousing, healthcare, biotech, pharmaceutical, electronics, government, transportation, and delivery.

Rush Tracking Systems

Drawing on its real-world experience with multiple vendors, applications, and frequencies, this Kansas City–based full-service RFID solutions provider specializes in inventory management, asset tracking, and supply chain management. Services include strategic impact analysis, hardware/software evaluations, systems integration, and installation.

Saft America

Based in Valdese, North Carolina, Saft manufactures primary and rechargeable lithium battery chemistries for the manufacturing, warehousing, and distribution industries. Saft's lithium technologies include lithium thionyl chloride for metering, tracking, and security applications, and the high-powered medium prismatic lithium ion for gas detection, computer backup, and the professional video market.

SAMSys Technologies

This Durham, North Carolina, company designs and deploys passive RFID readers (LF, HF, and UHF) that support EPCglobal standards and other tag protocols and frequencies. SAMSys also offers RFID consulting services, working with producers and suppliers of RFID tag products to facilitate the integration of current (and future) tag devices.

SATO America Inc.

SATO America has been manufacturing industrial bar code printers for more than two decades. For companies affected by short- and long-term RFID requirements, including its large installed customer base of Fortune 500 companies, SATO offers a one-step print and encode solution that the company claims is the fastest and most effective approach for a user to adopt RFID and comply with current mandates. Building on past experience with 13.56 MHz RFID technology, SATO has been involved at the forefront of RFID applications, working directly with Wal-Mart and its top suppliers, with the DoD, and with the leading RFID adopters in Europe.

Scan Solutions Inc.

Specializing in the recruitment and placement of direct sales, marketing, and technical professionals, Scan Solutions works with companies that provide software and other products for supply chain management, CRM, ERP, warehouse management, point of sale, and Auto-ID.

Secura Key

This Chatsworth, California, company produces the e∗Tag Quasar series of 13.56 MHz RFID reader/writers. Weather-resistant fixed readers and PC boards for embedded applications are available with RS-232, RS-485, TTL, and Wiegand communications. Secura Key also packages HF and UHF RFID inlays in cards, key fobs, pallet tags, and other form factors, with four-color graphics, barcodes, and other variable data.

Sentinel Business Solutions

Sentinel Business Solutions develops innovative data collection and RFID solutions. Services include software development tools and middleware platforms. In business for over three decades, Sentinel's in-house engineers, sales, and service professionals provide design, implementation, and support for sophisticated systems. They partner with many technology providers, such as Intermec, Symbol Technologies, and Zebra Technology.

Serviant Corporation

Headquartered in Newport Beach, California, Serviant specializes in enhancing organizations' supply chain logistics and asset tracking management through its RFID consulting, integration, and outsourced managed services.

Ship2Save.com

Ship2Save.com's inventory management systems can help a company manage warehouses and reduce shrinkage, and offer a number of tools for administering inventory. Its fleet management systems can track trucks, providing real-time location information that can help reduce the risk of theft of cargo on trucks and optimize fleet utilization.

Simply RFID

Simply RFID is a middleware provider based in Warrenton, Virginia. The company installs systems primarily for supply chain tracking and works in accordance with EPCglobal standards and specifications. Simply RFID works with end users for RFID applications from slap-and-ship to total integration.

SIRIT Technologies

Carrollton, Texas–based SIRIT Technologies has a Radio Frequency Solutions division that creates custom RFID readers that are embedded into industrial printers, hand-held computers, and cashless payment terminals from major manufacturers. The division also does research and development for smart shelves that can read RFID chips embedded into product packaging. The information can be used to alert employees when stock or inventory is low and to facilitate automated merchandising techniques for consumer products companies. Founded in 1993, SIRIT also has divisions that focus on Automatic Vehicle Identification applications, including electronic toll collection, parking and access control, airports, fleet, vehicle registration, and intermodal applications.

SkandSoft Technologies

SkandSoft Technologies' middleware offers intelligent two-way data, dynamic thresholding, event filtering, customized data-mining, interchange with industrial control systems, open-source methodologies, and variety of reporting structures. The company can integrate any vendor's RFID hardware and software applications, including proprietary platforms. Solutions include SCM/logistics, asset management, airline baggage tracking, point of sale, CRM, and transportation management.

SmartCode Corp.

SmartCode Corp. is a world leading EPC RFID manufacturer that has developed a revolutionary, patented, and cost effective EPC RFID mass production manufacturing technology. SmartCode Corp. is a member of EPCglobal and actively involved in shaping the next generation of low-cost EPC RFID Tags and Readers. SmartCode Corp. has developed the smallest EPC microchip.

SoftLogistics LLC

Softlogistics claims a profound understanding of the intricacies of RFID technology, as well as the processes of logistics and supply chain. The company has experience deploying nationwide RFID projects. Softlogistics provides turnkey RFID solutions. It can integrate its products with a customer's existing system or COTS products, and customize its products to fit each customer's specific needs.

Solzon Corp.

Solzon provides RFID middleware and integration services to manufacturing and distribution companies in the process industry. Its middleware products enable the capture of data at the point of activity and the deliver that data to the requisite applications within a company's ERP and warehouse management systems. Solzon's services are focused at configuring and integrating all the hardware and software components of the supply chain.

Symbol Technologies

Based in Holtsville, New York, Symbol is a global provider of wireless networking and information systems that allow users to access, capture, and transmit information at the point of activity over LANs, WANs, and the Internet. The company is developing handheld RFID readers that will work with its large installed base of bar code scanners.

Syscan

Syscan International develops and manufactures software and hardware to enable companies and public sector customers to operate efficiently in mobile computing and wireless applications. Products include ZFP-3, a suite of rugged in-vehicle and portable printers, and Systrak, a totally integrated business-efficiency solution focused on a highly sophisticated line of RFID components. Systrak solutions include TempaSure, Tireman, Meatrak, inventory management, readers, and smart card applications.

Sysgen

Sysgen specializes in developing solutions for the pharmaceutical, healthcare, and safety industries. Current solutions in place: amusement park ride inspections, elevator inspections, and fire marshal inspections. Sysgen offers consulting services, site surveys, beta lab build-outs, supply chain RFID systems integration and RFID solution development. Sysgen has developed various anticounterfeiting solutions for consumer goods companies and the pharmaceuticals industry.

Sysgen Data Ltd.

This Melville, New York, company's RFID installations include state government agencies, manufacturing firms, and the safety industry. Sysgen incorporates both stationary readers communicating in WANs and pocket PC-based technologies transmitting data wirelessly via 802.11.

T3Ci

T3Ci is an RFID analytics and applications company based in Mountain View, California, that develops and markets software and subscription services for users of RFID for major retail suppliers and pharmaceutical companies. T3Ci develops enterprise-class solutions, including inventory intelligence, brand protection, and order execution performance management.

Tacit Solutions Inc.

This Evansville, Indiana–based systems integrator has more than 16 years of experience in RFID, RF/Wireless, Auto-ID, mobile computing, and applications development. Tacit provides RFID lab testing for tag/antenna selection, placement, and portal design. The company also provides components and services including legacy-host interfacing, RFID middleware, shipping software, hardware and installation. Tacit's RFID applications cover manufacturing, supply chain, and Department of Defense environments.

TagMaster Inc.

TagMaster, Inc. is a global provider of long-range RFID technology offering high-performance products for demanding applications. TagMaster currently has more than 1,000 installations globally, often integrated with systems from one of its partners.

TagStone

TagStone is a provider of RFID solutions. It integrates RFID technology and software from leading manufacturers and developers into cost-effective solutions. The company works with global partners to offer its services almost anywhere in the world.

TAGSYS Inc.

With more than 15 years experience in implementing passive RFID systems operating at 13.56 MHz and UHF frequencies, TAGSYS specializes in item-level RFID tagging for tracking, tracing, and security applications across a spectrum of vertical markets and embedded systems. As a designer and manufacturer of complete RFID system solutions, TAGSYS provides chips, tags (read only, read/write and field-programmable multiread), readers (both fixed and handheld) and antennas, as well as consulting services. Included in the product line is the ARIO series, a low-cost 13.56 MHz tag designed to withstand harsh industrial environments, and the FOLIO flexible series tag that is designed for laminating between paper sheets and onto cardboard or plastic foils. Also offered is the MEDIO 13.56 MHz multiprotocol OEM reader series designed to be integrated with small devices. Antennas are available in a variety of configurations.

Tapemark

Tapemark offers a chipless implementation of RFID for brand protection and anticounterfeiting. Because Chipless ID uses a frequency well above conventional UHF, no chip is needed, significantly reducing the cost of the RF tag compared with that of traditional chip-based solutions. When used for anticounterfeiting measures, Chipless ID can validate product at points within the distribution channel. The Chipless ID reader can detect the presence of the Chipless ID tag, which uses microscopic antennas. For more secure applications, the tag's unique signature can be compared with a database. Tapemark believes that its Chipless ID system is ideal for companies that depend on revenue from consumables. For example, a copier manufacturer could equip its copier with a Chipless ID reader that allows the copier to operate only if the RFID signal shows that the manufacturer's own tagged toner cartridge is in place.

Techprint Inc.

This Lawrence, Massachusetts, company prints conductive inks in silkscreen for customized RFID and UHF antennas. It also custom-prints labels and identification products, including membrane switches and conductive circuits.

TEK Industries Inc.

Manchester, Connecticut–based TEK Industries designs and manufactures inexpensive custom and off-the-shelf hardware solutions for LF, HF, UHF applications, including customizable handheld and desktop RFID readers.

Texas Instruments

TI-RFID (Texas Instruments Radio Frequency Identification Solutions) offers a range of RFID and EPC transponder, reader, and antenna products for industrial, retail, and consumer applications. Texas Instruments is located in Plano Texas. Texas Instruments Radio Frequency Identification (TI-RFid™) Systems is an industry leader in radio frequency identification (RFID) technology and is the world's largest integrated manufacturer of RFID tags, RFID smart labels, and RFID reader systems. Approaching 500 million RFID tags manufactured, TI-RFid™ technology is used in a broad range of RFID applications worldwide including automotive, contact-less payments, laundry, livestock, pharmaceutical & healthcare, retail supply chain management, and ticketing.

TI is and active member of many standards bodies, including EPCglobal, ISO, ISO/IEC, ECMA International, ETSI, and national standardization bodies working to drive adoption of global standards of RFID.

Tharo Systems Inc.

Tharo Systems, based in Brunswick, Ohio, makes RFID label design software. Tharo is the author of EASYLABEL software for custom designing and printing product identification, bar code, and RFID labels. There are over 70,000 copies running around the world currently. Tharo has developed an RFID Wizard for its software that helps new users print and program HF and UHF smart labels.

The Kennedy Group

The Kennedy Group is a provider of integrated solutions for RFID applications. These include testing capabilities along with EPC-compliant antennas, printer applicators, installation, service, and support to help customers achieve compliance with outside mandates. Based in Willoughy, Ohio, the company provides manufacturing capacities in North America of smart labels and tags in 900 MHz and 13.56 MHz frequencies and offers an automated quality control inspection capability for EPC-compliant tags. Its printer/applicators can eject bad or quiet smart labels before application, to increase throughput.

Thin Battery Technologies

Thin Battery Technologies, based in Parma, Ohio, has purchased the technology to manufacture printed thin batteries (less than 1 mm thick) from Eveready Battery Company. Thin Battery Technologies (TBT) has more than 200 years of battery design and development experience on staff to provide integrators, converters and manufacturers with an environmentally friendly power source for microelectronic applications. Its battery designs are of a closed cell design and are not dependent on moisture or atmospheric conditions to generate power. Its electrochemistry is well suited for low temperature applications, such as data loggers and RFID applications for monitoring frozen foods.

ThingMagic

This Cambridge, Massachusetts, company says it invented the first agile reader for the Auto-ID Center in 2001. It has developed RFID solutions for a variety of target markets, cost requirements, and performance constraints. ThingMagic's RFID work includes RFID-enabling existing products, developing and deploying tag readers, and using RFID for product authentication.

Tompkins Associates

Tompkins Associates is a supply chain consultancy and systems integrator based in Raleigh, North Carolina. It works with clients on strategy and business case development, detailed implementation planning, product testing, custom software development, package software implementation, and integration of AIDC and RFID. It also provides distribution network design, DC and manufacturing facility design, transportation and logistics consulting, material handling integration, technology justification, selection, implementation, and integration. The company has a large integration lab in Orlando, Florida, where its clients can perform initial product testing and conference room pilot activities.

Topflight Corporation

This Glen Rock, Pennsylvania, company offers a variety of RFID label, tag, and card services, including printing, etching, insertion, coating, laminating, die cutting, serialization, and delivery in sheets or rolls. Numerous material and adhesive options are available, and tamper protection with device destruction can be designed in as well. Topflight can also print antennas in a range of designs and frequencies, such as a thin layer of copper on top of a seed layer of printed conductive ink, which offers greater performance at lower cost versus antennas printed with conductive ink only. Topflight is ISO 9001:2000 registered and manufactures products according to FDA cGMP procedures.

Toshiba TEC Corporation

TOSHIBA TEC is a manufacturer of retail and industrial information systems. As a total solution provider, TOSHIBA TEC, based in Japan, offers a complete package from consulting and system design to system installation, operation, and maintenance of electronic point of sale systems, electronic cash registers, electronic scales, barcode printers, RFID system, peripherals, and software information systems.

Tower Semiconductor

Established in 1993, Tower manufactures integrated circuits with geometries ranging from 0.18 to 1.0 micron. In addition to digital CMOS process technology, Tower offers advanced non-volatile memory solutions, and mixed-signal and CMOS image-sensor technologies. The Santa Clara, California–based company is utilizing its technical capabilities to offer reduced-cost RFID ICs. Tower also provides complementary technical services and design support.

Traxus Technologies, Inc.

This McLean, Virginia, company provides RFID consulting, implementation, and systems integration services to commercial and government customers. These services include technology evaluation, educational seminars, training, business-case development, pilot planning and deployment, production implementation, enterprise and supply chain integration, and business process modeling and reengineering. Traxus also has considerable experience in ERP systems.

TrenStar

Headquartered in Denver, Colorado, systems integrator TrenStar offers a pioneering RFID-enabled, "pay-per-use" model of mobile asset management. Its solution is designed to reduce transportation and operating costs for asset-intensive industries and companies invested in containers that move goods through the supply chain. In 1998, TrenStar UK (then known as KTP) formed a strategic R&D division to focus solely on RFID technology. Today, TrenStar is one of the few companies that has several successful, large-scale RFID implementations. TrenStar's clients include Coors, Kraft, Goodyear, and Burberry.

Tri-Star Consulting, Inc.

This Indiana-based systems integrator and VAR works in the RFID, bar-code, client/server, microcomputer, and networking sectors. It can also support client project teams by providing specific skills to each team.

Trivalent Solutions, Inc.

Based in Wilmette, Illinois, Trivalent's RFID solutions are designed to help manufacturers, distributors, and service organizations select, implement, and integrate RFID into their supply chains. The company's supply chain expertise and partnerships with RFID hardware suppliers allow for systems that mesh with a client's existing business processes. Trivalent's client base includes Fortune 1000 manufacturers, Department of Defense agencies, and mid-size businesses.

UPM Rafsec

This Finnish company develops and produces RFID tags for contactless smart cards and labels. The company has sales offices in the United States, France, the Netherlands, Japan, and Singapore. UPM Rafsec is part of the UPM-Kymmene Group, a producer of printing papers.

Vendor Managed Technologies, Inc.

Ann Arbor, Michigan–based Vendor Managed Technologies offers a data warehousing and reporting solution for retail suppliers to track RFID information down to the store level and integrate it with POS data provided by retailers.

VeriSign

Based in Dulles, Virginia, VeriSign delivers infrastructure services that make the Internet and telecommunications networks more intelligent, reliable, and secure. VeriSign has been selected by EPCglobal to operate the root Object Naming Service for the EPCglobal Network. The company is applying its networking expertise to a suite of managed services that allow trading partners to share RFID-enabled product data in real time.

ViVOtech Inc.

This Santa Clara, California–based company offers hardware and software for consumer contactless-payment systems. Payments can be made with an RFID-enabled credit or debit card, an RFID-enabled cellphone, a PDA or an access card at existing point-of-sale systems. ViVOtech's technology targets banks, retailers, and other wireless operators.

Vizional

Headquartered in Santa Monica, California, Vizional is an RFID product and services organization that specializes in large-scale supply chain visibility software. Vizional's Enable RFID product controls multiple networks of RFID tag readers, both wired and wireless, performing functions such as tag read/write, data filtering and the securing of tag data. The Asset RFID and Inventory RFID products allow customers to view and manage the flow of assets or inventory. Business rules can be defined to trigger real-time alerts so unexpected events don't become major supply chain problems. Vizional's Enterprise RFID allows customers to expand the functions of Asset RFID and Inventory RFID to multiple facilities or locations, and provides additional enterprise-level functions. The company also offers RFID assessment services.

Wave Data Technologies, LLC

Based in Rockford, Michigan, Wave Data Technologies is a systems integrator and technology consultancy that assesses, designs, develops, configures, pilots, deploys, and supports RFID-based solutions. At Wave Data Technologies, RFID solutions are aimed for any process where items or people need to be identified, tracked, secured, or moved.

Wavechain Consulting

Denver, Colorado–based Wavechain Consulting is an RFID middleware provider and systems integrator for small and mid-tier companies that want inventory tracking solutions. Utilizing open-source technologies, Wavechain provides passive RFID solutions that are compliance based.

Wavelet Technologies Inc.

Attleboro, Massachusetts–based Wavelet Technologies uses proprietary technology to design entire RFID tags, including antenna design, IC integration, and chip attachment. Wavelet Technologies tags can be integrated in packaging and have wide read ranges in challenging dielectric environments. Wavelet Technologies specializes in conductive inks and die cut or stamped metals (e.g., Al, Cu). The company conducts site surveys and product/tag simulation studies and uses EM SIM software to optimize RFID reader deployment. Wavelet Technologies constructs custom RFID test chambers that are fully FCC compliant, using optimized interior angle construction and ray analysis to minimize the interior volume (and cost) of a test facility.

Weber Marking Systems

Weber Marking Systems is a worldwide manufacturer and supplier of labeling and coding solutions, the most recent of which is its variety of RFID products and systems. Weber provides an assortment of RFID printers and printer-applicators that encode data onto RFID tags and labels, as well as produce printed bar codes, text, and graphics on those same smart labels. Weber also offers RFID label and tag material, plus RFID reject systems.

WhereNet

Santa Clara, California–based WhereNet provides real-time locating systems based on active RFID tags. The tags can be attached to cargo containers in a marine terminal, bins of parts on an assembly line, or vehicles in a yard. The company also provides software systems that enables users to quickly map their facility and pinpoint assets on the map.

Wild Mouse Software Inc.

Based in Hauppauge, New York, Wild Mouse provides software that operations personnel, IT, and system integrators can use to understand, design, and document distribution system requirements. Members of an in-house team can use the company's Mouse Map tool to define a model that includes physical layout and process information for a warehouse or light manufacturing facility. Once a model is developed, it serves as the backdrop for all communications on system deliverables. For system implementers, Mouse Map helps eliminate the late discovery of requirement gaps.

POINTS OF CONTACT

Accenture Corporation
RFID Practice Group
1345 Avenue of Americas
New York, NY 10105
(917) 452-4922

Alanco Technologies, Inc.
Technology Systems International, Inc.
15575 North 83rd Way, Suite 4
Scottsdale, AZ 85260
(480) 998-7700

Applied Digital Solutions Inc. (VeriChip)
1690 South Congress Avenue, Suite 200
Delray Beach, FL 33445
(561) 805-8001

Association for Identification & Mobility (AIM)
125 Warrendale-Bayne Road
Warrendale, PA 15086
(724) 934-4470

RFID-A Guide to Radio Frequency Identification, by V. Daniel Hunt, Albert Puglia, and Mike Puglia
Copyright © 2007 by Technology Research Corporation

California Department of Corrections
1515 S Street
Sacramento, CA 95814
Director's Office
(916) 445-7688

Department of Defense
Defense Logistics Agency
Automatic Identification Technology Office
Fort Belvoir, VA 22060
(703) 767-4012

EPCglobal US
Princeton Pike Corporate Center
1009 Lenox Drive, Suite 292
Lawrenceville, NJ 08648
Contact: Kelly Shearer
(606) 620-4671

International Business Machine
1133 Westchester Avenue
White Plains, NY 10604
(800) 426-4968

Ohio Department of Rehabilitation and Correction
Central Office
1050 Freeway Drive North
Columbus, OH 43229
Director's Office
(614) 752-1164

Technology Research Corporation
5716 Jonathan Mitchell Road
Fairfax Station, VA 22039
(703) 250-5136

Texas Instrument Incorporated
TI-RFID Product Information Center
6550 Chase Oaks Boulevard, MS 8470
Plano, TX 75023
(888) 937-6536

INDEX

RFID-A Guide to Radio Frequency Identification, by V. Daniel Hunt, Albert Puglia, and Mike Puglia
Copyright © 2007 by Technology Research Corporation